"No other construction professional has helped me shape my mental models about construction more than Todd Zabelle. Our sparring sessions over more than the last three decades have allowed me to see the shortcomings of the concepts that shape the management of our projects. They have made me seek better theories that enable project teams to produce high-performing building projects that make them proud and that improve our lives. They have made me a better researcher and teacher of construction. I encourage you to engage with Todd in the eleven sparring sessions the chapters in this book offer to change your mental models so that when it counts you will manage your projects with truly sound production concepts.

We have quickly gotten accustomed to seamless and transparent experiences when booking a trip, when banking online, or when getting from point A to point B. These experiences would not be possible without all elements of production being fully digital and connected. In the construction industry, we can only dream of such a simultaneous increase in customer value and efficiency of delivering this higher value. In fact, this will remain a dream unless we start to build our approach to managing projects on sound production management concepts that connect all elements of production and leverage the detailed production data that is now available from all kinds of sensing devices. Whether you are an owner, designer, or builder, *Built to Fail* equips you with the knowledge necessary for giving your construction project customers the kind of transparent and seamlessly integrated experience you have come to expect from your other service providers."

MARTIN FISCHER, PH.D.

Kumagai Professor of Engineering, Professor of Civil and Environmental Engineering, Stanford University; Director, Center for Integrated Facility Engineering; Senior Fellow, Precourt Institute for Energy

"It is news to no thoughtful person that the construction industry is broken. The challenge is understanding why it doesn't work so it can be fixed. Todd Zabelle's new book, *Built to Fail*, meets that challenge, but be prepared to shed some paradigms. Here are but two examples of Todd's advice: 'Get rid of baseline schedules' and 'Design is never complete.'"

GLENN BALLARD, PH.D.

Research Associate, Project Production Systems Laboratory (P2SL), University of California, Berkeley

"*Built to Fail* is a must-read for anyone investing in digital infrastructure and green energy projects in the future. It is critical information for anyone who wants to survive as a relevant investment option and combat the massive problems plaguing the industry. In *Built to Fail*, Todd Zabelle lays out a scientific and data-backed approach to deliver projects more efficiently and effectively. These methods are proven to reduce cost, use of cash, time, and risk while improving control of time to market. For owners or anyone in the industry, ignorance is no longer an acceptable excuse, or a luxury you can afford."

HUNTER NEWBY
Entrepreneur and Investor; Founder, Newby Ventures

"Don't read *Built to Fail* if you are satisfied with how the construction industry is performing. If you are dissatisfied, but are looking to solve the problem by 'collaborating better' in doing work in the same old way, don't read this book. However, if you are interested in learning a proven methodology that focuses on rethinking how construction work gets designed and made, then spend the time to read this book.

For over twenty-five years, I have witnessed and participated with Todd Zabelle in his efforts to bring operations science to the forefront of planning and executing construction projects. Todd is not an ivory tower theorist; rather a hands-dirty practitioner who has learned his craft by applying operations science to real projects and real work. He also has an unvarnished style that is upfront and refreshing. Todd's efforts and learning are well documented in this book. He presents a strong case for the industry to embrace project production management as the foundation of rethinking product and process design to improve cost, schedule, quality, and owner-value outcomes.

When you finish the book you may conclude that the remedy is 'simple'—getting back to basics. While it may be simple, as part of teams that have been working to implement these ideas over the past twenty years, I can assure you that it is not easy. Having the playbook that Todd provides in the book will certainly make it easier for you to chart your path and begin implementing these ideas. And Todd's candor and sense of humor will keep you smiling along the way.

Thanks, Todd, for providing the industry with the benefit of your hard-earned wisdom!"

WILL LICHTIG

Chief of Staff, Executive Vice President, The Boldt Company

BUILT TO FAIL

TODD R. ZABELLE

BUILT
TO FAIL

WHY CONSTRUCTION PROJECTS
TAKE SO LONG, COST TOO MUCH,
AND HOW TO FIX IT

Forbes | Books

Published by Forbes Books, Charleston, South Carolina.
An imprint of Advantage Media Group.

Forbes Books is a registered trademark, and the Forbes Books colophon is a trademark of Forbes Media, LLC.

Printed in the United States of America.

10 9 8 7 6 5 4 3 2 1

ISBN: 979-8-88750-158-1 (Hardcover)
ISBN: 979-8-88750-159-8 (eBook)

Library of Congress Control Number: 2023917974

Cover design by Matthew Morse.
Layout design by Megan Elger.

This custom publication is intended to provide accurate information and the opinions of the author in regard to the subject matter covered. It is sold with the understanding that the publisher, Forbes Books, is not engaged in rendering legal, financial, or professional services of any kind. If legal advice or other expert assistance is required, the reader is advised to seek the services of a competent professional.

Since 1917, Forbes has remained steadfast in its mission to serve as the defining voice of entrepreneurial capitalism. Forbes Books, launched in 2016 through a partnership with Advantage Media, furthers that aim by helping business and thought leaders bring their stories, passion, and knowledge to the forefront in custom books. Opinions expressed by Forbes Books authors are their own. To be considered for publication, please visit **books.Forbes.com**.

This book is dedicated to all the people that design, make, and build things, from engineers that design things to the people that make these designs reality, including carpenters, concrete finishers, electricians, iron workers, millwrights, pipe fitters, operating engineers, riggers, sheet metal workers, welders, and so on.

CONTENTS

VISIT OUR WEBSITE FOR ADDITIONAL GRAPHICS, AND LARGE-SCALE VERSIONS OF MANY GRAPHICS INCLUDED IN THE BOOK.

During the past century, we have created an approach to delivering capital projects that is built to fail.

The more we implement elements of the current approach, the worse project outcomes become. If some is good, more is not always better.

Beginning with Babbage and Taylor and later with the advent of construction management, we have created an approach based on managing administrative work rather than focusing on what really matters: how to design, make, and build things.

Only through resolving this "gap" will we be able to solve the project performance challenge.

ACKNOWLEDGMENTS

MY PROFESSIONAL CAREER would not be where it is today without some incredible people whom I would like to acknowledge. Martin Fischer, Stanford University, you are perhaps the person with the most impact on my professional career, providing thought leadership and introductions to leading thinkers around the world. Glenn Ballard, you are the master of conciseness with an incredible ability to simplify the most complex thoughts. James Choo, you are the mind reader, always thinking what someone is thinking before they even think it. Roberto Arbulu, the field marshal always in control of the situation and getting shit sorted. Mark Spearman, you are brilliant. And to my father, George Zabelle, who can design, make, diagnose, and repair anything—I mean anything. Simon Murry of acumen7; Aran Verling; Nigel Harper; Mike Robins; John Lohan; Bernard Dempsey and Ray O'Rourke at Laing O'Rourke; Mark Reynolds of Mace; Joe Gregory; Gary Fischer, Wayne Crabtree, and Jim Craig at Chevron; Jan Koeleman and Jim Banazak of McKinsey & Company, all of whom challenged us to clarify and simplify our thinking and push us to make our solutions better (even when we didn't want to hear it).

And a big thank-you to those who assisted with writing the book: Kristin Buettner, Dan Reuter, Amanda Homelvig, and Brian Green.

CHAPTER 1

A FLAT-EARTH MENTALITY
IN A WORLD THAT'S ROUND

BESIDES BEER AND BRATWURST, Germany is perhaps best known for its efficiency and engineering. *Boondoggle* is not the first word that comes to mind. But apparently no one told that to the people responsible for the German capital's shiny new public works project, Berlin Brandenburg Airport (BER), which opened eight years late (after multiple delays), thirty years after its initial conception, and came in at billions of dollars over budget. Even on the eve of its long-awaited opening, inspectors still found 120,000 defects throughout the facility.[1]

Go west across the North Sea and you get the UK's Crossrail, a transit megaproject being built under central London that rivals BER for boondoggle status, with its completion date pushed back several times and its costs exceeding initial funding allotment by

---

1   Sabine Kinkartz, "Berlin's New Airport Finally Opens: A Story of Failure and Embarrassment," Deutsche Welle, https://www.dw.com/en/berlins-new-airport-finally-opens-a-story-of-failure-and-embarrassment/a-55446329.

about $4 billion.[2] Finally, four years after its originally scheduled completion date, it will become fully operational in 2023—one hopes. "Next year, it'll definitely be ready, we swear" is a tune we've heard one too many times.

Here in America, we aren't doing any better. Examples abound. Consider, for example, the Purple Line metro extension in Washington, DC, a project so infuriating in its failure to launch that the contractor building it quit. As of this writing, it is four and a half years behind schedule and a few billion over budget, with no end in sight.[3]

Attempts at revitalizing the nuclear power industry, which many believe to be a critical element of energy transition, have been negated by significant cost and schedule overruns. The situation resulted in indictments, prison time, bankruptcy (McDermott and Westinghouse), and increased rates for customers. Using modular construction was supposed to revolutionize the industry by making it more cost effective and safer to build nuclear plants. "Modular construction has not worked out to be the solution that the utilities promised," said Robert B. Baker, an energy lawyer at Freeman Mathis & Gary LLP in Atlanta and former member of the Georgia Public Service Commission, the state utility authority.

Those problems have led to an estimated $13 billion in cost overruns and left in doubt the future of the two plants, one in Georgia and another in South Carolina. Overwhelmed by the costs of construction, Westinghouse filed for bankruptcy on March 29, 2017, while its corporate parent, Japan's Toshiba Corp., was forced to divest strategic business units to avoid bankruptcy. The problems extend

---

2   "Crossrail: Elizabeth Line Due to Open on 24 May," BBC.com, May 4, 2022, https://www.bbc.com/news/uk-england-london-61095510.

3   Katherine Shaver, "Maryland Purple Line Construction Will Resume in August, Officials Say," *Washington Post*, May 6, 2022, https://www.washingtonpost.com/transportation/2022/05/06/purple-line-maryland-construction-start/.

beyond Westinghouse, as France's Areva needed to be restructured, in part due to delays and huge cost overruns at a project in Finland. The miscalculations underscore the difficulties facing a global industry that aims to build about 160 reactors and is expected to generate around $740 billion in sales of equipment in services in the coming decade, according to nuclear industry trade groups. Upgrades to the Panama Canal completed in 2016, Southern Company's Kemper County coal gasification plant, Wembley Stadium, and the list of project cost and schedule overruns seems almost infinite.

IN SPITE OF OUR BEST EFFORTS CAPITAL PROJECTS ARE NOT GETTING ACCEPTABLE RESULTS			
31%	**48%**	**21%**	**9%**
did not meet cost or schedule	met either cost or schedule	met both cost and schedule	met cost, schedule, production attainment

Data from offshore oil and gas projects.
Source: Independent Project Analysis (IPA)

At the same time, contractors undertake way too much risk for the return on investment they realize. Every contractor, no matter how large, is only one step away from bankruptcy. Over the past several decades, numerous contractors have failed or have avoided failure through being acquired, including Guy F. Atkinson (now part of Clark Construction Group), Morrison–Knudsen (Washington Group, then URS, then AECOM), the collapse of Carillion, the acquisition of AMEC by Wood Group, and the UK's mighty Laing, which O'Rourke purchased for £1 after three challenging projects. A few years back, the giant Korean shipyards that serve the energy sector also found themselves in need of restructuring.

Founded in 2015, Katerra planned to revolutionize the construction industry using advanced production processes and technologies to manufacture building assemblies and systems. Even an over $2 billion investment by Softbank and others could not keep the start-up from filing for bankruptcy in June 2021. Katerra was well funded and run by seasoned executives (some I know personally), who enjoyed a very successful career in other industries including technology and energy. These are very smart executives who know what they are doing.

All this is occurring with the construction industry being a major consumer of energy and a contributor of carbon and other types of waste.

The construction industry is as big as it is important. Without construction, our society would cease to be. We would have no ports, roads, or bridges. No cell towers, power grids, or water grids. No hospitals. No schools. No homes. Regardless of whether you live in the city, the suburbs, or the country, probably not a day goes by when you don't interface with or benefit from some major capital works project. Virtually every human being on the planet is affected by the industry.

And the opportunities for growth are enormous, and our capacity to complete projects must keep up with the global demand for buildings, facilities, and infrastructure. One recent study revealed that between now and 2060, global building floor area will double, and just the amount of building floor space required will be equivalent to building an

> The construction industry is as big as it is important. Without construction, our society would cease to be.

entire New York City every month for the next forty years. According to the International Energy Agency (IEA), an estimated $100 trillion plus will be required over the next three decades to reach net-zero carbon emissions by 2050.

However, for all its size and importance, the industry is deeply, disastrously broken. Project after project comes in behind schedule and over budget. The scale of waste is staggering: $1.6 trillion each year gone because of delays or cost overruns.[4] Ninety-eight percent of megaprojects—the kind of vital infrastructural projects humanity needs to live, work, and thrive—suffer from egregiously blown schedules and busted budgets. Projects are tied up in a byzantine snake pit of administration, bureaucracy, and fragmentation. It is amazing that anything gets built at all.

To put it bluntly: if the people running construction were instead supervising the hospital delivery room, a lot of babies would never make it into the world.

This is not just bad for business. It's bad for humanity. Behind every building project are real-life stakes that affect real people. Capital projects are feats of human ingenuity and engineering that make modern life worth living. Dams control the flow of waterways and provide clean drinking water and hydroelectric power. Airports, highways, and train terminals facilitate the passage of innumerable people from one place to another and grease the wheels of global commerce. Power plants and other types of energy infrastructure keep the lights on, the factories churning, and society functioning smoothly.

---

4   F. Barbosa, J. Woetzel, J. Mischke et al., "Reinventing Construction: A Route to Higher Productivity," McKinsey Global Institute, February 1, 2023, https://www.mckinsey.com/capabilities/operations/our-insights/reinventing-construction-through-a-productivity-revolution.

When they don't get built, we all suffer. And we all pay for it, in one way or another.

How did the industry become so dysfunctional? What are the root causes? And what can we do about them?

The problem is complex and multifaceted, but it revolves around one major blind spot: the industry fails to see projects as production systems that can be managed according to the principles of operations management.

The stakeholders have sidelined operations management as a marginal field that is relevant only during the postdelivery phase of a project. This is incorrect. Operations management, i.e., management of operations, is the answer to the fundamental question, "How do we actually get *XYZ* built?"

Putting operations management front and center in the process (rather than as an afterthought) violates the construction orthodoxy and its holy gospel: the Project Management Body of Knowledge, or PMBOK™. The Project Management Institute (PMI) has hundreds of thousands of members around the world, and whether you know it or not, if you're in construction your work is heavily influenced by this book. PMBOK addresses topics such as time and cost management, project scope, quality, and risk and procurement management. Though these functions are important, they are by no means the only activities needed to bring a project to completion. In reality and from a lean perspective, you may not need these activities at all, but if it is a new facility, you *will* definitely need to design, construct, and commission it!

Organizations such as the Association for the Advancement of Cost Engineering (AACE) build upon the PMBOK, providing detailed specifications for how to perform cost estimating, forecasting, and reporting.

PMI regards operations management as a knowledge set outside of formal project management, something that comes into play after the asset is delivered. My argument, in contrast, is that it is the knowledge to actually deliver a project, because you're taking inputs to make outputs. I'll elaborate on that in the chapters to come, but for now, suffice it to say that this is important because it represents a huge gap in the current way of thinking and doing. The utility of operations management in designing and building—production, in other words—has been given short shrift. To the detriment of literally everyone.

Ignoring operations management is so deeply rooted in industry practice in large part because the way we do things relies too much on outmoded, late nineteenth- and early twentieth-century approaches to management and production.

Broadly, we can separate the field of capital project delivery into three eras: Era 1, Era 2, and Era 3.

Era 1 was the era of Frederick Taylor and scientific management. Taylor attempted to revolutionize industrial efficiency and labor productivity, by separating *planning* from *doing* and rigorously measuring and tweaking each stage of the production process to boost efficiency. In the early twentieth century, Daniel Hauer applied Taylorism to construction. What did not work in manufacturing is also causing confusion in construction. The separation of planning and doing means that the two sides are fundamentally disconnected, out of sync. You need a synthesis and synergy of planning and doing to ensure projects are not just built but built well, on time, and on budget.

Era 1 is by no means over—construction companies and the people running them are still stuck on Taylor/Hauer. While other industries have evolved, we are spinning our wheels.

Era 2, the era of predictability, took shape in the late 1950s and 1960s and was preoccupied with the predictability of project outcomes. This was—and remains—the era of project management; I say *remains* because, like Era 1, Era 2 still dictates how things are planned and built (or not built, more accurately.) Era 2 focuses on the administrative aspects of managing a project rather than designing and building. Consequently, how you do work, how you design something, make it, deliver it, install it, etc. is given short shrift. Instead, industry stakeholders fixate on how we organize the project, track the project, and motivate people.

The result is that lawyers, planners, risk managers, schedulers, and other administrative functions are getting in the way of architects, engineers, and craftspeople—you know, the folks who design, make, and build the things. This problem manifests itself in that reviled but all-too-common practice of purposeless meetings, or "meetings about meetings." There's a funny story in the book *Offshore Pioneers: Brown & Root and the History of Offshore Oil and Gas* about spending two weeks getting ready for the meeting, one week in the meeting, and two more weeks recapping the meeting to spend another two weeks getting ready for the next meeting. Meanwhile, no one is actually talking about the vital technical aspects—how to build the damn thing.

While I was having dinner with Frank Gehry, Frank explained how an owner was hiring a project management firm at a fee of 8 percent to protect themselves from potential cost overruns estimated to be 8 percent. It is nonsense.

Another symptom of this dysfunction is severe fragmentation. The vertical span between the owner who needs an asset built and the people doing the design, building, and engineering work is vast, separated by many, many layers. There is a hierarchy of owners, architects, construction management firms, and specialty contractors. Sometimes hierarchy can be a source of productivity and efficiency—not so in this case. Instead, you have a complex system of intertwined layers with competing interests that don't communicate well with each other and often have mismatched objectives and agendas.

Horizontally, the fragmentation means that a given project requires a dizzying multitude of independent companies, contractors, subcontractors, and craftspeople, instead of all these functions consolidated under one roof. If you've ever tried to do any work on your house, you've probably encountered something similar yourself. You need to hire a dozen different people (painters, electricians, a guy who does flooring, a mason, and on and on). Well, this fragmentation is as true of billion-dollar capital projects as it is for what should be a simple kitchen renovation.

The ability to effectively integrate and coordinate all this outstretches the capabilities of the methods and process used today.

So all of us suffer from the shortsightedness, the inefficiency, the strangling bureaucracy, and the lack of vision of Era 1 and Era 2. The solution is what I call Era 3.

Era 3 is the era of profitability where the focus shifts from administration to production. It really began in the mid-'90s, when a few of us started breaking from the dominant models and introducing new ways of working. Era 3 is the hope for the future—but we're far from realizing that vision.

That vision is this: rather than focus on administration, we're going to focus on production. Stop filling out forms, worrying about contracts, and writing each other emails, and concentrate on designing, making, and building things.

The vision of Era 3 also calls for embracing the litany of exciting technologies we have at our disposal: autonomous vehicles, robotics, high-speed networks, massive computing power, and emerging technology such as artificial intelligence (AI) and Internet of Things (IoT), etc. These technologies have a place in construction, as they do in other industries. The Era 1/Era 2 preoccupation with administration, contracting, risk analysis, and all those ancillary tasks causes the industry to ignore modern ways of getting things done. Technology could solve a lot of the logjams that make capital projects unnecessarily late and absurdly expensive. Many of the administrative tasks, for example, can be automated. Or rather than dispatching an army of planners to predict when materials will arrive or playing telephone tag for an hour, we can use GPS tracking, IoT sensors, and automated scheduling updates to keep us apprised in real time of where things are, where they're going, and when they'll get to where they need to be. The solutions are already there. We just need to implement them.

The war I'm waging is really a two-front one. On the first front, I'm trying to get the stakeholders of the industry to even just recognize the value of operations management. That will require a lot of drilling through habituated thinking to make the case that we must evolve beyond Era 1 and Era 2 and move decisively into Era 3.

The second front is a question of "mental shifting," because even people who are receptive to these ideas are skeptical that we can implement them in practice. Not because the ideas are flawed but because everything is so ossified: stuck in its ways, resistant to change. Every time I present at a conference, people come up to me after and say, "Yeah, I hear you. I wish we could do it differently. But how do we get people to change?"

The answer, which I'll discuss later in the book, won't be found on the project site or in the board room, and certainly not in management schools. Rather, it's right there in your own head. If we want to change our business, production, or commercial models, we must first shift our mental models. Everything emerges from there. In the final chapter, we will explore a paradigm-shifting framework explaining how to think and use the gray matter between our ears.

WHAT IS THE VALUE, AND WHY SHOULD YOU CARE?

I get it. If you're running a construction company—or any number of the industries that depend on construction to build its plants, factories, data centers, facilities, and clinics, not to mention the roads, railways, bridges, and airports used by everyone—you're thinking, "Why does this matter? Why is it valuable?"

We can describe various "value buckets" that the lessons within this book will offer. You can increase your share price. You can outperform your competition. You can control time to market. You can reduce the duration and cost of your capital projects while mitigating carbon emission, improving quality, and reducing risk. All these have a real impact on your bottom line.

Let's start by thinking about different types of entities out there. Service providers are mostly human-based enterprises (accountants, attorneys, creative advertising, software), working in commercial buildings and sometimes at home (post-COVID).

A producer of products, in contrast, needs heavy assets—processing plants; manufacturing facilities; logistic assets; ships, planes, and railways and their facilities—to produce the goods they sell.

And there's a third category we'll call civil infrastructure: roads, bridges, dams, etc.

It's the latter two that I am primarily looking to reach. Leaders in those fields understand that the ability to effectively use your capital determines how investors assess your work. Investors seek a return.

That return depends on what you do with that cash. (For civil infrastructure, the return on investment is quantified differently, as the investor is the government and the return is the benefit to the general public. But the basic principle remains the same. The input must yield an output. Its benefits must exceed its costs.)

The efficacy of investment decisions and development yield influences net present value, return on capital employed, and other metrics used to quantify how well companies are using their capital. For producer-type companies (the capital-intensive businesses), the return is heavily dictated by the ability to control time to market and deliver the right products to market at the right time. The slowness and general unpredictability of capital projects (which these producer companies rely on) impede these goals. So the ability to control time to market, depending on the industry and situation, bears on how to think through or frame your projects.

If you're not prudent in how you deploy cash in a project, you end up with unfavorable outcomes, which are reflected in the development yield. Net present value and return on capital employed projects, or even projects that are rendered unfeasible when actually done through the approach outlined in this book, become very doable. What does this mean? You can do more projects and drive more revenue and profit.

The ability to deliver a project using *less* cash and within the optimal time dramatically improves the value created by the project.

You want to expend capital in the least amount and over the shortest period prior to generating revenue. In other words, cash outlay should be short (in time) and minimal (in dollars).

For capital-intensive industries, such as companies building or maintaining digital infrastructure, manufacturing facilities, basic materials, and energy companies, this matters a great deal. How do I most effectively deploy the capital and control time to market?

The same is true for life sciences, which contend with their own set of challenges around coordinating R&D, the patent process, regulatory approvals, and production.

> **The ability to deliver a project using *less* cash and within the optimal time dramatically improves the value created by the project.**

Generally, if you're a life sciences company, you launch R&D, file for a patent, and then apply for the requisite permits from the Food and Drug Administration (FDA). Only then can you really think about the construction or tooling you need to produce the product. But the patent application process, and the approval process with the FDA or whichever body oversees your work, can take a very long time.

Tooling and construction are also time intensive, and also unpredictable. So do you do that in advance or wait for approval? It's a huge risk, because a delay or rejection by the regulatory authorities means that your construction or tooling investment could go to waste, having invested a ton of cash in production that now won't even happen.

This is not just a theoretical risk. I've seen it occur. A company we were in talks with went through the whole drill: they started construction *before* they won approval. But their petition was rejected by the FDA—so they wasted a huge amount of resources.

They made a bet—and lost.

But the other alternative is equally risky: you can wait on approval, which might take years, before starting construction. That just postpones the point when you can actually start earning a return on your investment. And if you're working with patented intellectual property, it means you've squandered a few years off a valuable patent that is only valid for ten or twenty years.

Likewise, in data centers, they must constantly think about how to match demand with capacity. If you build too much data-center capacity (exceeding demand) then you've tied-up cash that isn't really creating value. And if you build too little that can't meet existing demand, you lose out on potential revenue.

How do you match demand and capacity in the most economically beneficial way? You need to be able to control time to market—to control the delivery of the data center being built.

All this has to do with the duration of the project. The project delivery *is* the opportunity, and within the delivery process there exists a hidden opportunity that changes everything. It makes unfeasible projects feasible, reduces carbon, decreases cash outlay, mitigates risk, and most importantly enables predictable realization of project objectives. Perhaps, most importantly, it reduces stress.

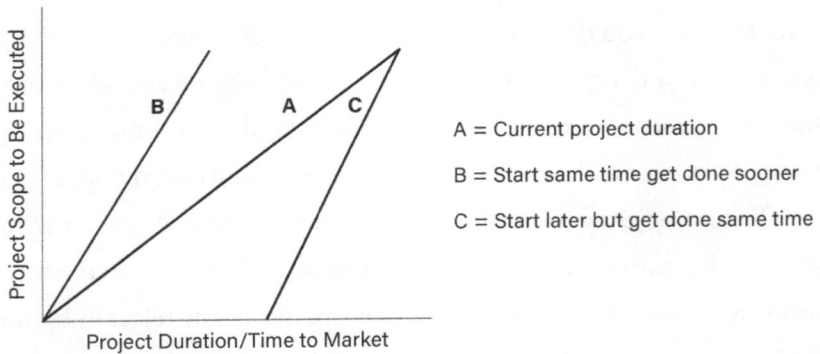

A = Current project duration

B = Start same time get done sooner

C = Start later but get done same time

y-axis: Project Scope to Be Executed
x-axis: Project Duration/Time to Market

Whether it concerns data centers or a pharma plant, because there are so many unknowns, it is extremely difficult to find the sweet spot between *too early* and *too late*. The ideas and techniques my company has perfected—which I'll discuss in this book—offer the key for breaking out of this maddening guesswork.

It's all about controlling time—and time is money. Moreover, work in *process* (a production management term) is not to be confused with work in *progress* (a project management term). We'll talk about this throughout the book too. The ability to manage time to market *through effective and predictable delivery of capital projects* is key.

When you learn to see the gap, you will understand that the cost, use of cash, and schedule duration for any project anywhere can be significantly decreased. It is no longer a technical problem but rather a mental challenge.

Benefit of Controlling WIP During Project Delivery

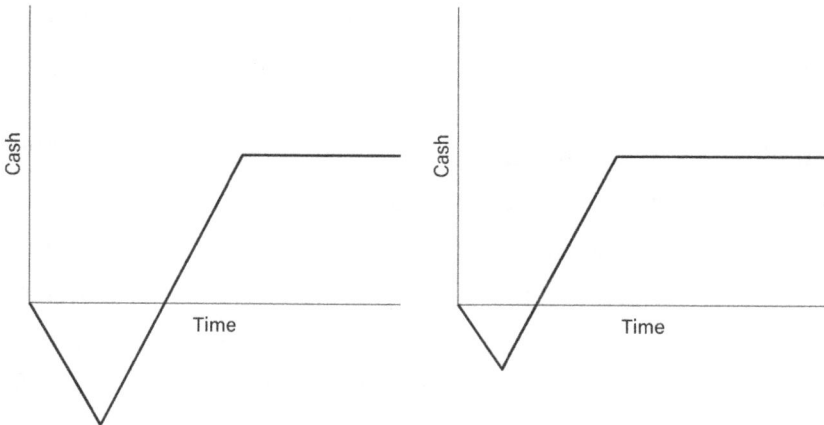

ACTUALLY, THE EARTH IS ROUND

For as long as I can remember, I've had a passion for designing, making, and building things. When, as a high school student, someone told me I could get paid to work out and get a tan, I figured construction seemed the logical choice for summer work. I've spent my entire working life in the industry, and I have been investigating these problems and devising solutions for the better part of forty years. I've collaborated with some of the largest infrastructure enterprises in the world and have authored numerous papers on engineering, fabrication, and construction. Having been born in San Francisco and grown up on the San Francisco Peninsula, I was there when Silicon Valley came to be and grew to what it is today. Silicon Valley is truly a unique place when it comes to innovation and the people that drive innovation. The people, the conversations, and the culture are like nowhere else in the world. Perhaps most relevant to this book are the technical contributions made to the construction industry by Stanford University and UC Berkeley, with which I have had the opportunity to collaborate for decades.

I have had the opportunity to learn from and collaborate with recognized leading thinkers in the industry, including Martin Fischer and Boyd Paulson of Stanford University; Glenn Ballard, Phil Kaminsky, and Iris Tommelein of UC Berkeley; Greg Howell, Simon Murry, Mark Reynolds (now CEO of Mace), and James Choo, Roberto Arbulu, and Mark Spearman of Strategic Project Solutions (SPS); and leading scholars outside the industry such as Anil Seth. At SPS, we have developed an educational program that brings leading experts in their field, such as Terry Slattery, to educate our team. Equally important is serving customers who are recognized as the best in the world, customers who drive us every day to create better solutions for their businesses, including Aran Verling; Nigel Harper; John Lohan; and Mike Robins, formerly of Laing O'Rourke; Gary Fischer, retired from Chevron and now executive director of the Project Production Institute (PPI); Wayne Crabtree, also retired from Chevron; and Dave McKay, retired from Hess.

Beginning with Heathrow Terminal 5, my company has provided project production management (PPM) solutions to customers and projects ranging in value from $500K to over $55 billion. These projects include massive infrastructure and industrial projects globally as well as aerospace and defense programs.

That said, you may have never heard of me, but you will most likely recognize many of the initiatives and innovations that I have been involved with. I was most proud when Construction Users Roundtable (CURT), in the fall 2019 edition of *The Voice Magazine* described three important trends occurring in construction: advanced work packaging (AWP), lean, and PPM, all of which I have been involved with in some way or another.

My tenure as organization leader goes back to 1993, when I founded my specialty construction company. At the time we were one of two companies in the world implementing the Last Planner system (the other was Brown & Root) and at the same time we were innovating the use of 3D and 4D modeling to optimize detailed construction operations in partnership with Martin Fischer. In 1997, I was one of the founders and owners of the Lean Construction Institute (prior to making it a not-for-profit in 2000), an organization that works to reform production management in design, engineering, and construction. If you are an advocate or user of AWP and its predecessor, workface planning (WFP), you may find it interesting to know that the framework for WFP referenced a paper I coauthored with James Choo, Iris Tommelein, and Glenn Ballard in 1998.

In 1998, my construction company Pacific Contracting was featured in the UK government's Rethinking Construction report, sponsored by the deputy prime minister, and in 2000, I founded my current firm, Strategic Project Solutions, that has deployed PPM solutions in support of complex and critical projects for national and integrated energy companies, independent exploration and production companies, hyperscale data center and other technology companies, aerospace and defense contractors, shipyards, heavy fabrication shops, and owners and builders of major civil infrastructure, among many more, and often in partnership with the world's leading consulting firms. In 2012, I founded the Project Production Institute (PPI). Through PPI, we have partnerships with Cal Poly, San Luis Obispo; Stanford University; and Texas A&M University to educate people in the concepts laid out in this book.

If you are involved in the general building sector and have had the opportunity to work for Sutter Health or for any other client that has been influenced by Sutter's development of its lean construction framework, you'll be interested to know I was there with Norm Barnes introducing these ideas to a few Sutter people, including Will Lichtig, who also went on to make a major impact in the industry through his drafting of the Integrated Form of Agreement championed by the Lean Construction Institute (LCI) and Sutter.

Focused on how to make more money as a contractor, my journey of learning began by reading numerous books on project management (big complex books written for the aerospace industry) and later thinking, as many still do today, that it is about motivating workers to work harder and, therefore, attending seminars on all sorts of stuff related to motivation. But once I began to learn about production and its underlying science of operations, I could see the problem and what to do to fix it. This is a journey many will make, starting with how to use project management frameworks more effectively, then moving to the people element, then how to become lean, and then finally arriving at the production perspective.

I've been a consumer of construction services at times, so I know this business, inside and out, top to bottom, at every level.

And at every level, there is something seriously wrong, to the point that the highly litigious nature of the construction industry constrains engineers from considering and using new materials, as well as preventing building department officials from accepting new technologies. To watch construction companies pursuing industrialized construction using the same old materials, including wallboard and the associated tape, mud, and paint, is nothing short of astonishing. But the cost and risk of alternative products constrain innovation.

Perhaps my greatest contribution has been the translation of operations science for construction projects and the associated supply chains.

As arrogant as it may sound, I challenge anyone to provide a situation that you believe cannot be optimized through the concepts outlined in this book.

The divorcing of planning from doing, and the preoccupation of administration, contracts, and scheduling over designing and building has had a disastrous effect. It's like we're in the time of the ancient Greeks, sailing a ship across the waters, and we're lost at sea. Instead of worrying about our dwindling provisions, the captain and the first and second mates are desperately trying to figure out how to avoid sailing off the edge. Those of us who understand the problem, and how to fix it, are shouting, "Look, there is no edge! The world is round! Why don't we deal with the fact that we're running out of food and water?" And these pleas fall on the deaf ears of mariners who don't know what they're doing or where they're going.

This book is written for anyone who is a producer or consumer of engineering and construction services, particularly senior-level professionals who recognize the problem and are eager for a resolution. The industry's dysfunction is troubling, but it can be fixed. These cost overruns, schedule delays, and project carnage don't need to happen. There's a better way. And I'm going to show you how.

CHAPTER 2

ERA 1: THE ERA OF PRODUCTIVITY

"**I JUST WANT TO KNOW**—how did things get so bad?" asked Joe, an executive with a large energy company that is a customer of my company. We were talking shop over dinner. Joe was in charge of capital projects, an important role considering that the global energy conglomerate has tens of billions of dollars tied up in major construction works around the world: offshore oil and gas platforms, liquefied natural gas plants, chemical plants, refineries, pipelines, and so forth. When it comes to building major assets, few organizations are as engaged as Joe's company.

In appearance, Joe is kind of how you'd expect a man in his role to look: a middle-aged corporate guy with a slight Texas drawl. I hadn't known Joe for long, but I admired him because it was clear that he just wants to do what's right. However, that desire only amplified his frustration over the glacial progress and bureaucratic gridlock that marred some of the capital projects he was overseeing, as they were plagued with a litany of compounding problems: production delays, tied-up cash, mounting legal claims. And some of their contractors were teetering on bankruptcy.

Why was it so hard even for a company such as Joe's, a powerful and well-run company with virtually bottomless resources and unparalleled organizational and logistic capacity, to build the vital infrastructure it needed to keep its company—and the millions of people and countless businesses who depend on the energy it provides—running?

How did we get here? he wanted to know. Why are all these projects so bad?

Joe's company had made a bet on megaprojects. Their strategy was to use the latest construction means and methods possible, employ the latest thinking, and do what none of its competitors, or seemingly no one in any industry, could do: plan and complete a project of significant scale on time and on budget.

But what they thought was the "latest thinking" was actually just familiar Era 1 and Era 2 mistakes: the warmed-over leftovers of the industrial era and the clumsy attempt by the construction industry to apply nineteenth- and twentieth-century methods of production and industrial organization.

Era 1 is defined by the advent of centralized planning/project controls, the separation of planning from doing, and the rise of bureaucracy. It was, and remains, an era of brute-force work from craft workers and calling that productivity. And it's a big reason why the industry is so dysfunctional.

> **Era 1 is defined by the advent of centralized planning/project controls, the separation of planning from doing, and the rise of bureaucracy.**

KEY THINKERS OF ERA 1— THE PANTHEON OF PRODUCTIVITY EXPERTS

Charles Babbage

In 1832, if you were a factory owner in rapidly industrializing England, you might sit down in your leather chair by the fire, stuff some tobacco in your pipe, and crack open a heavy tome titled *On the Economy of Machinery and Manufactures*—about as fun as it sounds. Nevertheless, Babbage's (1791–1871) thinking changed industrial production forever.

The most important principle of the economics of manufacturing is the division of labor among the persons who perform the work. In his studies of what occurred on the factory floor, Babbage found that high-skilled, higher-paid workers were spending (wasting) part of their day on low-skill tasks. This might seem obvious now, but what is clear to us today was not necessarily evident then. In any event, thanks to Babbage's work, managers, owners, and a newly emerging class of industrial managerial specialists launched a movement to reshape labor by having each worker specialize in what they do best.

As the Industrial Revolution progressed in the nineteenth century, the master-craft guy, who was an expert in painstakingly building some custom-made thing in his shop, was replaced by high-volume mechanical production. Then came specialization. And the legacy of this complicates things.

Today, construction is still very craft based. Compared to other industries, it's less automated. You drive down the street, you see guys working on a house—most of it is craft. Humans applying their trade. Sure, the hammer has evolved into a nail gun, but there's still an individual—a craftsperson—manipulating that tool.

Specialization has led to fragmentation. As the various craft functions have been spun off, they start working at cross-purposes with each other.

It is said that back in the day, when a steel company was hired to build a bridge, they'd make the steel, fabricate the steel members, and erect the steel. But later they said, "We don't want to fabricate it; we just want to make steel. Let someone else cut it up." So now, just to handle the structural steel alone, you need to have a structural engineer, someone making steel, someone who fabricates the steel into steel members, someone who erects it, someone who does the shop drawings, and a hoisting company. What before required one company now involves five.

The specialization has led to fragmentation, and fragmentation is one of the millstones around the industry's neck and why things are so damn inefficient.

Frederick Taylor

If you work in an office, depending on your position within the hierarchy, you may have had, at one time or another, the misfortune of encountering that busybody who is studiously monitoring your work and how you spend your time, perhaps to assess your "efficiency," perhaps to assess your contribution to the firm down to the finest detail. Maybe this person is a superior, maybe an outside consultant. Some paper pusher with a clipboard, a ballpoint pen, and an air of haughty importance.

Well, Frederick Taylor (1856–1915) was an early iteration of that guy—and he literally wrote the book on it. His research revolutionized the way the captains of industry ran their shops, and his seminal work, *The Principles of Scientific Management*, birthed the field of the same name. Taylor was a mechanical engineer who identified the inefficiencies in how manufacturing plants operated and in particular how workers did their work. In the quest to optimize production and maximize efficiency, armed with a stopwatch, he dedicated his professional life to observing laborers in action and breaking down their labor process into the minutest of elements—*minute* in the sense of "extremely small" as well as in the temporal sense. Taylor's great contribution to industrial management was time studies: How long do things take, and how can they be altered to shave off a few seconds or minutes from each part of the process to boost productivity?

The consequence of Taylor's popular theories of management science and industrial engineering was the separation of planning from doing—which in practice means white-collar guys telling blue-collar guys what to do, which is problematic because in practice it is mostly the blue-collar guys who do the actual building and know what is going on, what needs to be done, and how to do it.

For example, one of Taylor's contributions was the concept of "functional foremanship." Concluding that the role of a factory foreman required myriad responsibilities and functions, he reasoned that no one person could do the job. So he separated these functions into four planning foremen and four production foremen.

Planning foremen included a route clerk (who laid out the sequence of operations), an instruction card clerk (who would specify instructions for individual workers), a time and cost clerk (who managed the schedule and budget of the project), and a disciplinarian (the enforcer of rules and regulations on the shop floor).

Production foremen included a speed boss (responsible for ensuring workers completed their tasks on time and avoiding production delays), a gang boss (overseeing materials, tools, and machines), a repair boss (the maintenance and repair specialist), and an inspector (who monitored the quality of output).

Some of these titles might seem antiquated or of little relevance to a modern-day work environment, but look closely and you can see in this nineteenth-century approach to industrial management the structure of modern-day organization of construction: a clear delineation between planning and production. While perhaps efficacious in theory, it's been a disaster in practice, and we've been stuck in a Taylorist rut ever since.

Taylor was also an advocate of piece work instead of paying an hourly wage: for every shovel you make, you get ten cents. In construction today, we have the unit rate contract; contractors get paid for how much they produce. This, too, is a legacy of Taylor, whose fingerprints are all over the place.

Daniel Hauer

Taylor was so influential in his time that others took the baton from Taylor and built on his work in a way that also reverberates through the construction industry today.

One of those individuals was Daniel Hauer, who translated the principles of scientific management to construction. His magnum opus, *Modern Management Applied to Construction*, was published in 1918 and is now in the public domain, so it's easily accessible to anyone who wants to page through its two-hundred-plus pages and see for themselves why the industry has gone astray. It's unlikely you've heard of this book, much less read it. But if you

In construction today, we have the unit rate contract; contractors get paid for how much they produce. This, too, is a legacy of Taylor, whose fingerprints are all over the place.

earn your living in the construction industry, you are almost certainly influenced by its ideas.

Much of what we do today, in terms of trying to get things built, is an exact replica of what Taylor and Hauer did in their time.

Henry Gantt

Henry Gantt worked for Taylor and made his own lasting contributions to the emergent field of scientific management. Construction professionals best know Gantt for his eponymous chart: a bar chart that illustrates the duration (start to finish) of various tasks in a project. The horizontal axis represents time; the vertical axis displays tasks. Besides simply organizing a project and ensuring timeliness, Gantt's intent, like Taylor's, was to figure out scientific, systematic ways of getting more work out of each individual worker. The problem with how bar charts are used today is twofold: (1) it now reinforces the separation of planning from doing, because the people creating the charts are too far removed from the actual work on-site, and (2) it does not depict the time work is waiting to be processed (what, in the world of operations science, is termed *queue time*—later in the book you will learn why this is so important).

Furthermore, a bar chart does not depict the work in process (or WIP) building up during the delivery of a capital project. This is important as our analysis concludes the ratio of raw process time (the time actually doing the work) is often a small fraction of the total cycle time (the time it takes for an item to flow through the required process). And often, cycle time is a fraction of lead time. Using a process flow diagram such as the one that follows allows us to see the WIP including where it is building up. The majority of the time is related to queuing or waiting rather than production.

Service providers and product suppliers have learned to use WIP to shield their labor from upstream variability and to enable optimization of their capacity. But what is good for them locally is disastrous for the owner of the facility. As the time between operations increases, so does the duration of the schedule and the cost of the project. This is the hidden opportunity—manage the WIP.

Frank and Lillian Gilbreth

If you have seen the movie *Cheaper by the Dozen*, then you know the husband-and-wife duo of Frank and Lillian Gilbreth and their work in the area of motion studies.

The Gilbreths studied workers in search of the optimal way to perform work from a kinetic perspective. Building on the theories of Taylor, the Gilbreths observed various workers engaged in various tasks, quantifying the amount of time each could be performed to systematize production. Frank Gilbreth first studied bricklayers at his own contracting business, and Lillian Gilbreth published a book teaching homemakers how to apply motion study to minimize wasted energy.

Since their work did not take into context the end-to-end process, much like Taylor and Hauer, their thinking can be problematic because, again, Era 1's preoccupation with squeezing the maximum labor out of every worker is misguided and myopic. But the industry is still stuck on this concept because that's the way it's been done for so long.

As you will learn later in the book, reducing WIP and the associated queue time is a significant opportunity. The following diagram depicts how time or schedule duration gets extended for the owner as each contractor looks to use inventory to optimize their productivity.

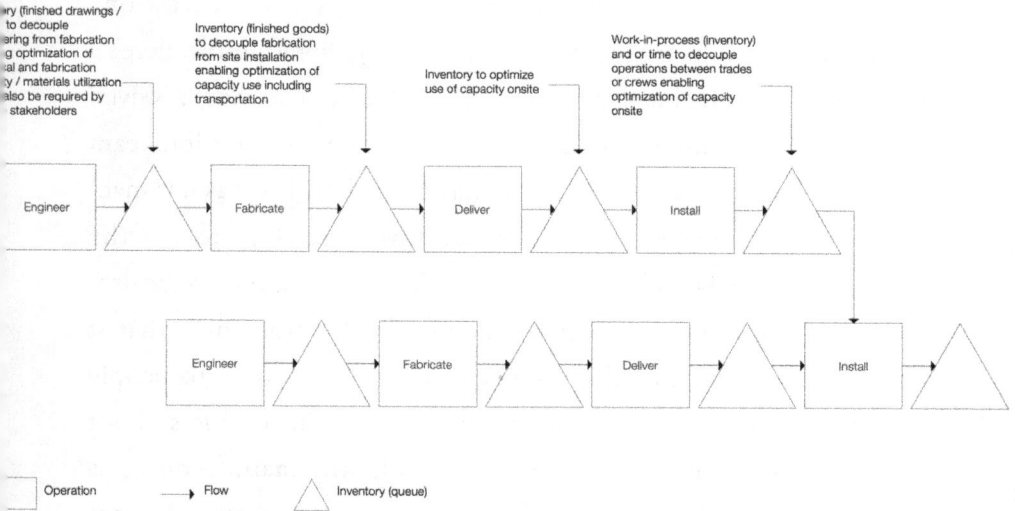

Engineer → Fabricate → Deliver → Install

Engineer → Fabricate → Deliver → Install

☐ Operation → Flow △ Inventory (queue)

Project Production Institute

BEHAVIORAL SCIENCE

Era 1 was also the era of adapting behavioral psychology (an emergent field during the early twentieth century) to the workplace. Instead of focusing on dividing labor to maximize efficiency and standing around with a stopwatch as workers load pig iron, the question became "How can we motivate workers by looking at industrial production through the lens of human needs (beyond the basic physical needs)? What is the relationship between those needs and productivity?"

While Schmidt (the daft, fictional laborer Taylor described in *The Principles of Scientific Management*) is out there shoveling coal as Taylor is doing time studies on him, the Gilbreths are monitoring his motion and figuring out a better way to shovel, and the behaviorists are asking how to connect the interests of owners with his needs—not really to improve his life per se, but to make him shovel faster.

Today, much of what goes on in construction is traced to this emphasis on behavior and behavioral psychology. Attend an industry conference and you are likely to find on the program at least one session devoted to mental health and well-being. For example, here's a description of a scheduled talk at a recent conference I attended: "Given the bevy of modern-day pressures that impact human condition, team members can slip into burnout and malaise ..." And that is not inaccurate. Work pressure, stress, overwork, these are all factors of the workplace that adversely affect both productivity and mental health. I'm not saying they shouldn't be addressed. The problem is that it becomes a kind of closed loop where the focus changes to people rather than process. In manufacturing, in contrast, it is less about people and more about the process, which is why manufacturing is better at getting things done, while those of us in construction are constantly fuming over our projects coming in late and over budget. All of which is self-inflicted.

A DIFFERENT PATH

The auto industry emerged during the time of Taylor and the start of the Era 1 practices we struggle with today. Henry Ford in the US, Frank G. Woollard in the UK, and, later, Kiichiro Toyoda and Taiichi Ohno of Toyota Corporation were all on a quest to perfect production of motor vehicles by answering the question, "How do we make work flow through a process?" Not "How do we squeeze workers by dictating the exact way and number of times per minute they should strike a nail with a hammer?" but "How do we get more out of the process?" As a matter of fact, both Ford and Toyoda were adamant that workers earn a fair wage and that the way to improve efficiency was through a process lens and not chasing worker productivity.

This differing path results in way more efficiency and productivity, as seen in many of the analyses comparing nonfarm industries to the construction industry. A while back we overlaid two productivity graphs, one from the National Bureau of Economic Research (NBER) and another from Paul Teicholz of Stanford University, on a timeline of innovations in both manufacturing and construction. What we observed was very telling: (1) for the most part, beginning in the early 1900s, the construction industry has shown *no* material improvement in productivity, and (2) other industries have shown significant gains in productivity.

The net impact over the forty-eight-year period shows that manufacturing productivity ended up about 3.6 times higher than construction productivity at the end of the period.

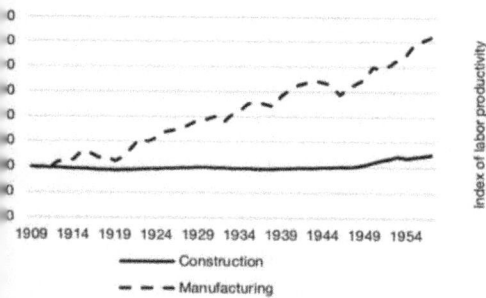

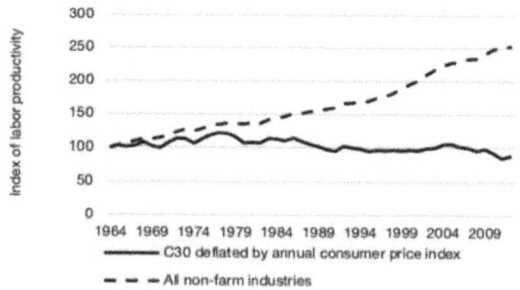

If you have a process that is sequential in nature—*A* follows from *B* follows from *C*—and if you maximize how much each worker can do, it may actually take longer if you put each one at max production, because there is variation in how much the *A* worker, the *B* worker, and the *C* worker can produce. The respective tasks are different, and the workers themselves are different (and produce at different rates).

So even if the workers at *A*, *B*, and *C* are going at max speed, you're not optimizing the process, because in a sequence, the final output is only as fast as the slowest worker—the old adage about the chain being only as strong as the weakest link. There's a bottleneck, and you're building up inventory at that point in the chain. More on the technical aspects of this later.

Imagine you're working the kitchen at a fast-food burger chain. Everyone making burgers is working at max speed. The guy forming patties works at max speed; and the guy cooking them works at max speed; and the guy putting the lettuce, tomato, and ketchup on the bun works at max speed. But if it takes one minute to form a patty, thirty seconds to place the lettuce/tomato/ketchup on the bun, and three minutes to cook the meat, guess what happens?

For those of us of a certain age, we remember the MacDonald's commercial that promoted Big Macs. "Two all-beef patties, special sauce, lettuce, cheese, pickles, onions on a sesame-seed bun." Radio listeners would win a Big Mac belt buckle for reciting the ingredients correctly in a minimum amount of time. But then Burger King came along with "Have it your way." What people didn't realize is that Burger King changed the configuration of the production system, moving the inventory from finished goods (burgers made) to raw materials or ingredients that could be cooked and assembled to order. This enhanced customer service, improved quality, and reduced cost.

Taylor did not actually account for any of this; to him, the more work you could get out of a worker, the better, without regard for what others on the line were doing. Ford and Woollard reasoned that we can only get production through the processes as fast as the slowest part of the chain will allow, while Toyota revolutionized production globally by building upon Ford's foundation. But for construction, even with the lean construction movement, this is still a major blind spot.

THE RISE OF BUREAUCRACY

While the productivity consultants were looking to optimize operations and the behavioral scientists were working to understand what motivates workers, another group was working on how best to structure an organization. Researchers such as Fayol, Weber, and Chandler studied the challenges of how to scale the organization. But Era 1 thinking was still in play.

The folly of Era 1 thinking is plainly evident at a very large liquid natural gas (LNG) project, built in Australia.

One of the things they did is they bought all the materials in advance and shipped them all to a laydown (a yard). We're talking billions of dollars of materials and equipment. And it just sits there. At one point the lead partner had $100 billion in construction projects underway. With all the steel and pipes, vessels equipment, generators, precast elements stacked up at various sites, they had to draw on their line of credit to pay shareholders their dividends. And no one said, "Maybe it's not a good thing that we've got billions in shit laying around in projects we don't even need for a year or two and have warranties expiring." This project, along with others doing the same, resulted in the resignation of the CEO.

Besides the accumulation of inventory, other problems were evident that conveyed how frustrating Era 1 practices can be. A simple task like building a light stand required five or six different trades: a guy to dig a little hole for the foundation, another guy to put in rebar, another guy to place the concrete, another guy to put in the steel, another guy to install the lighting apparatus. But that guy needs bolts, and they're not in the work package the planner put together, so he can't even do the work. Why is this the case? Well, because the planners, in their work packages, dictated it. And no one even asked the obvious question: "Why build a light stand when the light could instead be bolted to the massive steel structure itself?" It is the separation of planning from doing and the fragmentation into multiple trades transforming a task that should be relatively simple into becoming inordinately complex.

This LNG project is a manifestation of all the Era 1 mistakes: orchestrated by several of the biggest companies in the world, with ample resources made available to ensure its completion—and it still doesn't go right. The hundreds of planners and schedulers are out of touch with guys on the ground. They have consultants out there doing time studies; they're using a Gantt chart to do the planning; it's highly craft based; they've got psychologists on-site preoccupied with behavior and mental health, which is not a problem in and of itself but misses the forest for the trees by assuming the problem is the humans rather than the process.

Thanks to Era 1, what was once relatively simple has become complex. Taylor has separated planning from doing (you need both an electrical engineer and an electrician to handle a single function). Babbage transformed craft production into manufacturing through division of labor and cleaved trades from disciplines. Gantt charts, a.k.a. bar charts, emerged to figure out the sequence of each task in a project. The Gilbreths introduced motion studies.

Thanks to Era 1, what was once relatively simple has become complex.

And finally, you've got a battalion of consultants using new theories in industrial/scientific management and psychology to assess how to motivate workers by connecting their inner needs with the task at hand. Not to mention all the managers supervise them.

All these new theoretical/administrative/managerial/consultative fields give rise to a new class of professionals—a bureaucracy unto itself, wherein "How do we organize the project?" displaces the previous focus on "How do we build this thing?"

Large engineering and construction companies emerge, housing these professionals under one roof, but the functions, despite their appeals to efficiency and productivity, don't necessarily add value to the customer. They end up doing what bureaucracies do best: work for work's sake. Administration, accounting, compliance, etc. Rather than focusing on designing and making things, they focus on how to structure the company. This sets the stage for Era 2, the "era of predictability," when the preoccupation with administrative functions becomes even worse. That's the subject of the following chapter.

CHAPTER 3

ERA 2: THE ERA OF PREDICTABILITY

AS THINGS PROGRESSED into the midtwentieth century, the thinking of Era 1 was supplemented (not supplanted) by what I call Era 2, which, like Era 1, has never actually ended and still inhibits construction from being an efficient and functional industry that gets things done on time and on budget.

Era 2 was/is the era of predictability of capital project delivery. New project management methodologies such as the critical path method (CPM) and program evaluation and review technique (PERT) were developed. Later, earned value management (EVM), the phase-gate process, and advanced work packaging (AWP)/workface planning (WFP) were added to the mix. Together, these approaches are recognized by PMBOK as valid protocols for managing a project. But the preoccupation with project scope, time and cost management, quality, risk, and procurement management once again misses the mark. While these factors are important, other activities are vital for ensuring a project is completed successfully. Therein lies the industry's giant blind spot.

Era 2 reinforced the separation of planning and doing that had begun in Era 1. Starting in the post–World War II era, the United States Department of Defense (DOD) needed to provide annual reporting to Congress to account for funding appropriations for their various programs. In pursuit of this objective, they developed (in concert with Booz Allen) PERT. PERT is a project planning tool that visualizes a project's timeline. DuPont, in partnership with Remington Rand UNIVAC, later developed the critical path method, sometimes called *critical path scheduling*. As time went on, CPM has become the standard method for creating and managing construction project schedules globally.

> ## Era 2 reinforced the separation of planning and doing that had begun in Era 1.

CRITICAL PATH METHOD

Remington Rand was looking for applications for their UNIVAC mainframe computers, while DuPont was developing an approach to figure out how to optimize investment between cost and time—something they first started exploring during the Manhattan Project. The foundational text of CPM was published as a paper at the Eastern Joint Computer Conference in 1959 in a journal article titled "Critical-Path Planning and Scheduling," by James Kelley and Morgan Walker. In this paper, the authors laid out four reasons for CPM:

1. to form a basis for prediction and planning,

2. to evaluate alternative plans for accomplishing the objective,

3. to check progress against current plans and objectives, and

4. to form a basis for obtaining the facts so that decisions can be made and the job can be done.

The authors posited an inverse relationship between project cost and duration. If you add more people or resources, it'll go faster but cost more. If you want it to cost less, you must use fewer resources. That is the time-cost trade-off that later evolves into the time-cost-quality/scope trade-off, a.k.a. the iron triangle: you can have any two but not all three. This is a well-known "law" in a range of fields, not just in construction. As a matter of fact, during a visit to Cabo, I once overheard an interior decorator explaining to their client they could have quality, low cost, or the work done sooner, but not all three—only two!

In practice, however, it is wrong, in part because it fails to account for fundamental operations science. I'll examine this in greater depth later in the book.

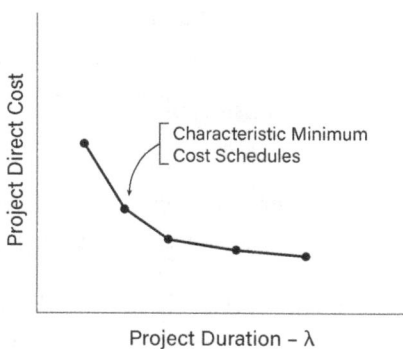

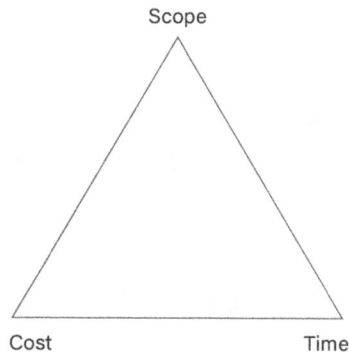

Project Direct Cost

Characteristic Minimum
Cost Schedules

Project Duration – λ

J. Kelley and M. Walker
"Critical-Path Planning and Scheduling"
1959 Proceedings of the Eastern Joint
Computer Conference

Scope

Cost   Time

Project Management "Triple Constraint"
aka "Iron Triangle"

The US Navy's Bureau of Yards and Docks then decided it wanted to adapt CPM, so it turned to Stanford University and offered a grant to civil engineer and faculty member John Fondahl. Fondahl published an influential (and lengthy) paper titled "A Non-Computer Approach to the Critical Path Method for the Construction Industry" in 1962. The title aptly summarizes the thesis: Fondahl's goals were "to present a noncomputer method for obtaining the benefits of critical path scheduling that is practical to apply to [construction projects]" and to "develop the possibilities inherent in a step by step, manual solution to overcome some of the shortcomings of computer programmed solutions."[5] Later, Fondahl expanded on this text in his "Methods for Extending the Range of Non-Computer Critical Path Applications," published in 1964.

The partnership between Fondahl and the navy was a fruitful one, and its impact extended throughout the construction industry as a whole. "Renewed for eight years, this pioneering support for research on construction work processes and organizations eventually covered many critical topics: application of operations research techniques to construction operations, development of time-lapse motion picture techniques, application of engineering economics to key decisions concerning construction equipment, and extension of the critical path method of scheduling construction operations."[6]

Shortly thereafter, the DOD and other agencies in the federal government started to mandate the use of critical path scheduling. That cemented CPM's place in the pantheon of project management, where it has remained since. Various software programs were developed for the mainframe, then the minicomputer, and finally the

---

5   John W. Fondahl, "A Non-Computer Approach to the Critical Path Method for the Construction Industry," Bureau of Yards and Docks, U.S. Navy/Stanford University, 1962, p. 3.

6   "Memorial Resolution, John W. Fondahl (1924-2008)," Stanford University, May 20, 2009, https://exhibits.stanford.edu/stanford-senate/catalog/zr928kb6805.

microcomputer, where today Oracle's Primavera is by far the most widely used CPM application in the world.

Later, EVM emerged as another method. In simplified terms, EVM basically says, "If we estimated the project to cost ten bucks, and we've spent eight, we're 80 percent done." What we've spent tells us what our progress is.

In the world of operations science, inventory is a proxy for time: you predict time by measuring how much work is underway in the process. Think about earned value—you get credit for spending more money because higher expenditures signify a higher rate of project completion. How does that affect inventory? This incentivizes spending as much money as possible. The more you spend and the faster you spend it, the better, because that shows progress.

This has been the modus operandi in construction since the 1960s, but it's all fundamentally flawed. We've built the house on a faulty foundation.

These new methods were codified in the PMBOK by PMI—and in the AACE Body of Knowledge and have become gospel, essentially.

In contrast, in manufacturing, especially in the auto industry, the focus was not on productivity per worker, but on the question "How can we make things flow through the process?" PMI, meanwhile, is fixated on how to predict when things will happen. This is achieved by creating a baseline schedule and budget and measuring to them. But to measure progress the plan can't be changed, because if they do change it, they can't measure progress to plan or variance off plan. Simply put, if they want to measure progress against a baseline schedule, they can't change the baseline. What do they do? Measure progress with an inaccurate schedule—or change the schedule? But then they can't measure progress against baseline schedule.

The AACE establishes the need for corrective action but, other than revisions to the plan, does not provide any insight into how to implement corrective action. This is the world of operations/production management.

In 1991, Fondahl pointed out[7] that all schedules become resource constrained: you can only go as fast as the resources you have. So ultimately what matters is managing resources. But no one listened to him when he said this.

So the almighty schedule was placed on a pedestal in the dogged pursuit of perfect predictability. But schedules often hurt as much as they helped, because they were to some extent divorced from the reality on the ground. Violations of a schedule whose mandates were infeasible if not impossible to comply with led to disputes. What begins to happen because of the requirements of CPM/CPS is that claims start to pile up. The attorneys swoop in. People bicker, often bitterly, over the schedule: Party A complains that Party B didn't do *XYZ* by this date as mandated by the schedule, but Party B retorts that Party A failed to make a certain decision that was necessary for enabling Party B to fulfill their obligations.

> **The almighty schedule was placed on a pedestal in the dogged pursuit of perfect predictability.**

Big capital projects thus start to become a morass of claims and counterclaims originating around alleged violations of the almighty schedule, which instead of serving as the means of coordinating the project now becomes a source of contention. A whole cottage industry of claims consultants and attorneys

---

7   John Fondahl, "The Development of the Construction Engineer: Past Progress and Future Problems," *Journal of Construction Engineering and Management* 117, no. 3 (1991), https://doi.org/10.1061/(ASCE)0733-9364(1991)117:3(380).

springs up, further entrenching the problem, because once someone starts making money from something, they have a very good reason to perpetuate it, even if whatever "it" is does not serve the common interest. Risk avoidance supersedes common sense and the ability to build shit.

HELMUTH VON MOLTKE—HEAD OF PRUSSIAN AND GERMAN STAFF, 1858–1888

But a century earlier, Helmuth von Moltke provided the framework for managing project schedules describing the exact framework for planning military operations that project professionals should adopt.

"No plan of operations extends with certainty beyond the first encounter with the enemy's main strength. Only the layman sees in the course of a campaign a consistent execution of preconceived and highly detailed original concept pursued consistently to the end. Certainly, the commander and chief will keep his great objective continuously in mind, undisturbed by the vicissitudes of events. **But the path on which he hopes to reach it can never be firmly established in advance. Throughout the campaign he must make a series of decisions on the basis of conditions that cannot be foreseen.** The successive acts of war are thus not premeditated designs, but on the contrary are spontaneous acts guided by military measures. **Everything depends on penetrating the uncertainty of veiled situations to evaluate the facts, to clarify the unknown, to make decisions rapidly, and then to carry them out with strength and constancy.**"

As many will admit, a baseline project schedule is outdated as soon as it comes out of the printer.

It is interesting to consider that even though the US DOD teaches the concepts set forth by von Moltke to leadership, they do not adopt this framework for the delivery of their capital projects. The same is true of how the US DOD has moved from centralized command and control to a more distributed model of decision-making.

PHASE/STAGE GATES

Originally developed by the chemical industry, use of the phase or stage gate process has become commonplace among large industrial facility owners. Stage gates are checkpoints in the project life cycle that are supposed to ensure that projects are progressing as planned and that any risks or problems are identified and addressed. They provide decision points that evaluate whether a project should continue to the next stage or be terminated. Stage gates are used to monitor and control the progress of a project and ensure that it is completed on time and within budget and meets the objectives set out in the project plan. Final investment decision, or FID, is a common milestone that often requires the board of directors' approval.

Similarly, the American Institute of Architects have also set forth standard phases for the delivery of a capital project based on the following phases: schematic design, design development, construction documents, bidding, and construction administration. Prior to schematic design, the architect works with the owner to define requirements and translate them in preparation for the design process. In product development and industrial construction, this is often referred to as the *basis of design*.

Stage gates are the basis of the waterfall approach commonly used in software development. Waterfall is a liner approach to software development that follows a sequential process of design, development, testing, and deployment. In this approach, each phase is completed before the next begins, and the process flows in a single direction. This method is characterized by its rigid structure and strict adherence to the plan, which can make it difficult to adapt to changing requirements or unforeseen problems. It is often used for projects with well-defined requirements, where the scope and requirements are unlikely to change. In the early 2000s, a group came together to advocate for replacing the waterfall approach to software development, and today the agile method has become commonplace.

As stated above, stage gate processes are based on a sequential approach to executing the work. But design and engineering of a facility are not sequential but rather iterative. This is where the administrative objective of checking progress and budgets runs head-on into the reality of how work is executed. The concept ignores the effects of large process and transfer batches, including the increase in work in process and associated increase in schedule duration. But perhaps the most ironic element of the stage gate approach is the implications of long-lead items. The lead times (the time it takes from order to delivery) for an item are so long that orders must be placed well in advance of the final investment decision. The need to place orders for long-lead items results in design and engineering being sequenced in a manner that focuses on ordering of long leads rather than the natural progression of the work (and nobody asks why the lead times are so long and what can be done about it). More on how to address long-lead items later in this book.

THE RISE OF CONSTRUCTION MANAGEMENT

To address the risk imposed by Eras 1 and 2, general contractors shifted to be construction management firms.

Under the construction management system, the owner hires the entity that was formerly known as the general contractor and is now called the construction manager (CM). The CM, tasked with managing the process, acts as the owner's agent: they're an extension of the owner and see to the owner's interest. The CM takes the design information, creates packages for different specialty contractors who do the detailed engineering, fabrication, installation—the work—and have them bid the job back to the CM's firm. And the CM's firm works off a fee based on the cost of all that work.

As a result, the CMs have lost their understanding of actual work because their focus shifted: now they have become administrators rather than producers. Or, as some say, "construction companies became brokers or distributors of specialty contractors." They trade in relationships: relationships with owners, specialty contractors, suppliers, etc.

At the same time, large companies (DuPont, Dow, etc.) divest their world-class engineering groups that ran projects internally and operate on a skeleton crew because it's not core business. The general-contractors-turned-CMs take their cue and say, "We can fill that gap by becoming the agent of the owner." And they "lean down" because general contracting is no longer their core business either.

As that model matured and became status quo, universities followed suit, educating the next generation of construction professionals in the nascent field of construction management. What was once a curriculum that emphasized production was replaced by an education in administration. America's institutions of higher learning began—and continue—churning out future construction profession-als who only know administration: organizing the project, risk man-agement, procurement, schedule management, HR management, and other administrative things.

The book *Offshore Pioneers* chronicles the construction of the world's first out-of-sight-of-land offshore oil platform by Brown & Root. The story of this feat of engineering is also a story of the trends of Era 2. "Brown & Root's clients, the major operators in the North Sea, needed large, flexible management organizations to control and supervise the fabrication and installation of the concrete and steel mega-structures built in this era ... Vast, complex projects that mandated project managers to coordinate the actions of field operators, project contractors, and numerous subcontractors."[8]

However, the era of predictability had exploded the bureaucratic apparatus that drove these megaprojects. "Project management held within it a potential for duplication of functions, which could work against efficiency. 'I guess the best way to understand it,' recalled Jay Weidler, 'is that in the worst case you would have more people watching the work being done than those doing the work—more people keeping score than people actually producing things.' Brown & Root was not immune to ... what was known in the industry as the 'glorification of project management.' Huge North Sea projects during the boom years often required up to five years for comple-tion and included 2,000 to 3,000 personnel for project management functions."[9]

---

8   Joseph A. Pratt, Tyler Priest, Christopher J. Castaneda, *Offshore Pioneers* (Houston, TX: Gulf Publishing Company, 1997), 370.

9   Pratt, Priest, and Castaneda, *Offshore Pioneers*, 371.

The bureaucracy that first took root in Era 1 had now morphed into something bigger, more complex, and more cumbersome. Era 1 sparked Era 2, and both were here to stay.

Naturally, the people driving these changes touted them as a good thing: the next phase in the history of construction, a time when predictability—being able to pinpoint what would happen, where, when, and by whom—was the watchword. But what everyone forgot as Era 2 took root was the actual work—the production. The preoccupation with administration missed the forest (the work) for the trees (scheduling, HR, budgeting, risk management, and so forth).

> But what everyone forgot as Era 2 took root was the actual work—the production. The preoccupation with administration missed the forest for the trees.

A SELF-PERPETUATING PROBLEM

When my colleagues and I founded the Lean Construction Institute in 1997, we set out to change the game by focusing on production, and redefining control "from 'monitoring results' to 'making things happen,' with a measured and improved planning process to assure reliable workflow and predictable project outcomes."[10] But the industry is stubbornly resistant to change.

No one is denying the existence of the problem—the frequency of delays and the seeming unavoidability of cost overruns are self-evident. And at least in theory, no one in construction is content with

---

10   "LCI Tenets," Lean Construction Institute, https://leanconstruction.org/pages/about-us/lci-tenets/.

this reality, but it has become normalized pain. So why does this not translate into a collective will to change?

There are several reasons accounting for this stasis. For one, by this point, several generations have been educated and trained in construction management, and they don't know any other way of doing things. There's a kind of institutional inertia that's going to take years, if not decades, to unwind.

Second, the existing business models profit from Era 2 paradigms, whether it's the specialty contractors that actually perform the work, construction managers who oversee that, owners who profit from it, or the claims consultants and attorneys who run a lucrative trade dealing with the ceaseless legal wrangling. In any system, if the people running or operating that system profit from it, there's a powerful disincentive against shaking things up.

Furthermore, fragmentation exacerbates the situation and makes any kind of mass coordination difficult. If you need six different trades just to remodel your bathroom, the same crippling inefficiency applies to major capital projects, just on a larger scale. It's a lot harder to get twenty different entities to overhaul the way they do business than it is a single company or individual.

Underlying all this activity are mental models that keep the players of the industry stuck on a hamster wheel, running in circles over and over and over.

Basically, Era 2 has solidified certain industry structures. Like concrete, once it sets, it's very hard to break it apart and create something new.

An associate of mine—let's call him Bob—was venting to me recently about this situation. Bob is a mechanical contractor involved in the construction of a pharmaceutical plant. Per the standard Era 2 approach, the owner hired a construction management firm to administer the project, and the CM firm hired Bob as one of the contractors. Bob informed the CM firm that he needed to get something approved on his contract so he could buy materials to do required work. The CM firm dragged its feet in issuing that approval, thus leaving Bob's hands tied. His protests fell on deaf ears until finally the company got around to realizing that in order for them to do the concrete work then erect steel, Bob had to first install the underground piping. They needed to accede to his request.

But a rupture in one part of the process causes compounding delays further down the line. Consequently, because Bob wasn't able to put his pipes in, they couldn't put the footings in, and as a result of that, the erection of the steel was delayed too.

As things started getting jammed up, the young project engineer working for the CM kept referencing the schedule as he scolded, "You're late, Bob, why haven't you done this?" Bob was deluged with a series of nasty letters accusing him of violating the schedule—but there was nothing he could have done, because other people responsible for that schedule didn't do what they were obligated to do.

And as Bob pointed out in his defense, it didn't matter anyway because they hadn't even hired the steel guy yet.

It's a business model based on a catch-22, basically. The schedule becomes a system of logic unto itself, cleaved from the actual on-the-ground, on-site needs of production. This situation repeats at innumerable sites all over the world, every day: at the large LNG project mentioned previously, for example, where there were forty schedulers in the room trying to figure out what was going on on-site by interviewing the craft supervision (foremen and superintendents). At yet another very large civil project, a piece of software was developed to manipulate the schedule report being submitted from the CM to the owner. It's madness.

Another factor that perpetuates the situation is simply that planners, schedulers, and other administrative roles tend to earn significantly higher salaries than structural or mechanical engineers focused on production. The administrator becomes the person of value, not the technical person. At the request of an owner, we were involved in the recommendation of and perhaps even partnering with a CM firm for a specific project. The CM firm explained their business model and provided their hourly rate sheet. It was amazing to see licensed engineers billing out at $95–$110 an hour while the project controls and other administrative roles were over $180 per hour.

Even worse, there's a lot more at stake for, say, a structural engineer than a scheduler. If the former messes up, they could go to jail.

For the engineering firms, especially the big firms that are also CMs, they've made it so they can bill out the scheduler at a higher rate than the engineer, and they'd rather even "CM the engineering," which means outsourcing the engineering to others. They oversee it and collect hefty management fees. It basically makes them brokers, and they take their cut. Naturally, this balloons the overall cost of production, without adding any

> There's a lot more at stake for, say, a structural engineer than a scheduler. If the former messes up, they could go to jail.

actual value. The surplus is sucked up by the middleman. We are now observing specialty contractors that subcontract the work (in one instance an electrical contractor was subcontracting the electrical work!). Another layer becoming a construction management firm.

BURN AND EARN

One of the consequences of EVM has been a system of "burn and earn": burn more hours at a faster rate as an indicator of progress. But if what gets rewarded gets repeated, guess what happens? Some firms will front-load a project with the easiest work possible, burning a lot of hours at the start to maximize earnings. But after that, the work just gets increasingly difficult because they've burned the easy stuff.

I've seen projects that ran out of progressive work. There were no more hours to burn, but there was still more work that had to be done. Debate then ensued as to whether to call it nonprogressable or unprogressable work!

A project sets a budget for cost including the total man hours to do the work. Progress measurement against the budget is done monthly. Since the budget is the budget, the project runs out of or overruns the budget, meaning more work but no more budget. Since progress is based on spend, there is no more money or spend to progress. Glenn Ballard of UC Berkeley calls this eating dessert first.

When that happens, not only does the project hit an impasse, but burn and earn leads to—what else?—disputes underscored by threats of litigation. Then, between the owners and the contractors, they have to come up with the money. I saw one case in which a whole company—a large, prominent contractor—ended up in bankruptcy because burn and earn went awry.

I've seen other cases when the CM firm artificially drives up costs. In its work as agents of the owner, it takes the design package from the architects and engineers and create bid packages—packages of work that specialty contractors (concrete, electrical, mechanical, etc.) bid on. It uses a guaranteed max price (GMP) contract, and it tells the owner it'll do it for a fee of 5 percent and the most it can cost is, say, ten bucks. So it tries to get as near to the GMP as it can through various tactics, such as getting multiple bidders to include a crane in their bid. But when it comes time to do the work, the CM deducts the cost of the crane from each subcontractor and charges the owner for a shared crane. Meanwhile the CM convinces the owner to flip it to a fixed-sum job. So now the estimates come in, the job can be built for $9.27, and it says, "Let's just lock it in for ten bucks and we'll take it fixed sum," increasing its margin. Desperate for predictability, the owner eagerly agrees, only to find the risk is not cost but rather loss of revenue due to schedule delays.

PASSING THE BUCK

This is a typical scenario: the owner is trying to get a project built for some business purpose, typically one of the following: (1) they need to optimize their production capacity (e.g., they need another plant, or they need their existing plant to do better), (2) they want to bring a new product to market, and they need a new plant or they must modify the current one (e.g., they need to change the automotive assembly line to produce a new model of car), or (3) they must conform to a new regulatory requirement (e.g., they can no longer dispose of chemicals in the waterway or they must adhere to new emissions benchmarks).

The owner's need is to predictably deliver the project based on their budget, timeline, and operating requirements (for example, they need to produce X amount of product).

On the other side of the equation are the specialty contractors: tradespeople with expertise in mechanical, electrical, plumbing, structural steel, concrete, etc. They handle the detailed engineering—figuring out exactly how they're going to cut the beam, then buying the steel to do so, then cutting the beam, then welding the connections on, then shipping it out and erecting it.

They're trying to figure out how can they take someone's need for a plant and provide them with the steel frame.

In the middle between the owner and specialty contractor is the construction management firm. They get paid for providing the administration. They operate as brokers, bringing together owners in search of contractors with contractors in search of work. Someone needs an asset to do something, and there are guys who can enable that asset.

As middlemen, they naturally want a cut, the bigger the better, and to maximize their pay they engage in all manner of commercial contracting games to make the arrangement work to their advantage.

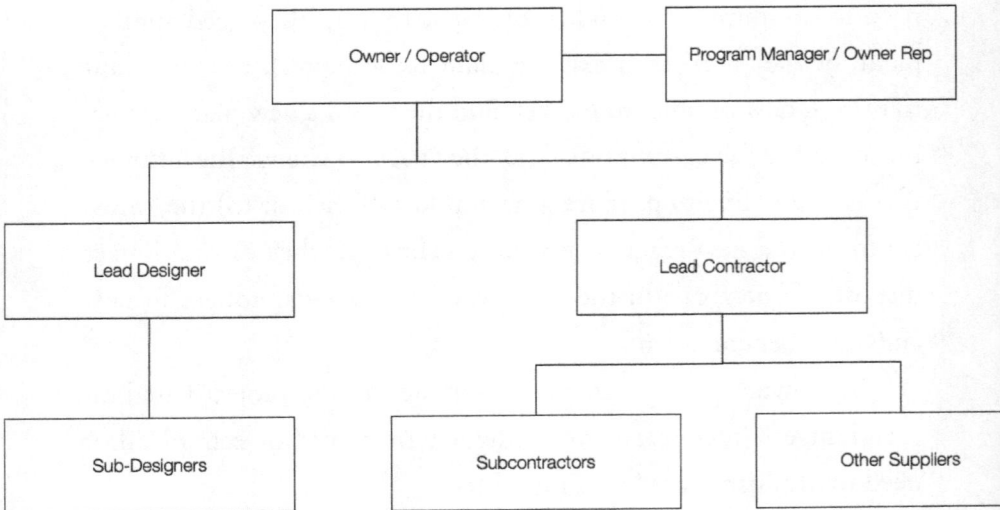

```
┌─────────────────────┐        ┌──────────────────────────────┐
│   Owner / Operator  │────────│ Program Manager / Owner Rep  │
└─────────────────────┘        └──────────────────────────────┘
        │
   ┌────┴───────────────────────────┐
┌──────────────┐              ┌──────────────────┐
│ Lead Designer│              │ Lead Contractor  │
└──────────────┘              └──────────────────┘
       │                    ┌────────┴────────┐
┌──────────────┐    ┌──────────────┐   ┌──────────────┐
│ Sub-Designers│    │Subcontractors│   │Other Suppliers│
└──────────────┘    └──────────────┘   └──────────────┘
```

Say you're building a house. The general contractor/CM might do a little work, but they won't paint, do plumbing or electrical, put in floors, or install the insulation. And there's much wheeling and dealing between CM and the craft guy. Because you probably won't get a price from the flooring guy; you'll get it from the CM.

Not only does the presence of the middleman administrator drive up costs, but the necessity of involving various trades, mediated by the CM, creates fragmentation, which adds to the complexity.

I'm the electrician. I give you a price for the job as long as I can start next Tuesday. But on Tuesday you're not ready for me to do my work. And I need to go to another job tomorrow. And because you're not ready, I don't get the money I expected—your delays are obstructing my ability to conduct business. But if tomorrow, when you're finally ready for me, I go to my other job as scheduled, that's also affecting your business, because now I have to finish this other job first before I can complete yours, which delays you further. Both of us are losing money as a result. Who's going to pay? This is the kind of situation that leads to everyone getting bogged down in claims.

Or you said you wanted to put in four electrical outlets and a switch, but when I arrived at the jobsite, you changed the work order to seven outlets and two switches. Now we have a problem.

Maybe the architect made a mistake and drew up plans that specified a need for only four outlets, but then the building inspector said no, you need one for every X number of feet, so you must install three more. These extra three are not accounted for in the budget or in the schedule. Who is going to pay for the extra three the electrician must install? The owner of the property? The architect because they should have known? The CM who is administering the whole thing? Or the electrical contractor because he should have known the right number?

The owner will say, "Fuck that, why should I pay more than what we agreed on because of someone else's oversight?" The architect will say, "No, my job is to design the home; how the hell should I know how many outlets are required by the building code?" It can't be the CM because they do not do any technical work. The electrical contractor will say, "That's not on me. Let the CM and the owner work it out."

And the CM is sitting on the outside saying, "It's definitely not my fault. And by the way, where's my cut?"

LOOKING BEYOND ERA 1 AND ERA 2—THE VISION OF ERA 3

The Lean Construction Institute got off to a strong start by focusing on production rather than project management, but as more people got involved, the original vision was diluted and the conversation morphed to the psychology and sociology of workers—how to make teams collaborate better, communicate better, and so on. Eventually, LCI reverted to exactly what the LCI charter stated we were trying to avoid: "a heavy focus on people and behavior versus understanding the fundamentals of production."

We are very different from other organizations because our first goal is to understand the underlying "physics" of production. This means we want to understand the effects of dependence and variation along supply and assembly chains. These physical issues are ignored in current practice and have no particular relationship to teamwork, communication, or contract. These human issues are at the top of a practitioner's lists of concerns because they do not, indeed cannot, see the source of their problems in physical production terms.[11]

Perhaps an AWP manual said it best when it proposed AWP offers increased productivity (Era 1) with enhanced predictability (Era 2)—this is truly the state we are in.

---

11   LCI formation document, 1997.

The current approach is described as Era 1 productivity with an overlay of Era 2 predictability.

But we have a chance to change course and revolutionize construction forever. What I call Era 3 is the alternative to Era 1 and Era 2. It's what we need to strive for. In some ways, we are already realizing it, but we have a long, long way to go. In the chapters that follow, I'll elaborate on this vision and lay out a plan to achieve it—a plan to finally break the cycle and free ourselves from the impediments of Era 1 and Era 2 forever.

The current approach is described as Era 1 productivity with an overlay of Era 2 predictability.

CHAPTER 4

INTRODUCTION TO ERA 3: THE ERA OF PROFITABILITY

IF NOT ADMINISTRATION, THEN WHAT?

That's the question that will help us forge a path out of the fixation of administration, bureaucracy, project management, and planning of Era 1 and Era 2 and show us a new way of doing things—doing being the key word. Era 3 is, in its simplest articulation, an era of profitability by way of making production the focus.

My argument is that projects are an assemblage of multiple production systems, which together yield a product, service, or result, and should be managed as such, drawing on operations science (OS) to understand and influence their behavior. Only through understanding basic production system types and behavior can we improve project outcomes. Now let's explore the world of production, including how it relates to construction projects and more importantly the profitability of your company.

PRODUCTION SYSTEMS

A production system is a specific or defined set of operations within a larger supply network or value chain that produces technical or physical output to satisfy an external demand. In 1979, Hayes and Wheelwright established that there are four primary "generic" production system types. These are *processing* (typically around pharmaceuticals, refining, chemicals—moving things through pipes), *connected line flow* (e.g., automotive—an assembly line where a combination of machines and people are putting things together), *disconnected flow/batch* (a line flow not connected by conveyance—examples include factories that make clothing, textiles, heavy equipment), and *job shops* (shops that design and/or make custom products where the first step is often design, and that also do machining, painting, and final assembly).

This is important because during the delivery of a capital project, all the above production system types come into play. But this, along with the attributes of each production system type, is not well understood from a technical perspective. How are they different? How are they best configured, controlled, and improved, etc.?

PROJECT PRODUCTION SYSTEMS

From an operations perspective, most of the focus of the current approach to project management is on the demand side, the schedule. What needs to be done by who, when. How it is done is completely missing from the equation. Understanding the various production systems, their specific challenges, and how best to configure and control them is critical to delivering a project successfully.

Therefore, the first step is to decipher between a schedule that in the world of operations is used to set the demand for the production system, and the production system that delivers to the demand. The production system being the process, operations, resources, control protocols, control mechanisms, and key performance indicators that take inputs and create the outputs necessary to deliver the asset in accordance with the schedule (to the extent possible).

This challenges the doctrine of PMI, which holds that (1) projects are undertaken to create a product, service, or result; (2) to do that you need resources: energy, labor, equipment, capital, etc.; but (3) the project does not provide a means for managing the resources from a production perspective. If you study construction management or even take a production/operations management course, that's what you're taught—managing projects is, for the most part, about administration.

Question it in class and you'll feel like Galileo explaining heliocentrism to the church elders. That's the orthodoxy.

Furthermore, most people in construction never think about the fact that those production systems already exist. They're operating. The A&E (architecture and engineering) firms; concrete companies; steel companies; companies that do inspection, engineering, installation, everything you need to build something—they're all in business. Owners and their construction management firms are basically plugging them together and integrating them to create an overall temporary production system: the project production system. That project production system is what will drive the cost, quality, risk,

> If you study construction management or even take a production/operations management course, that's what you're taught—managing projects is, for the most part, about administration.

time (schedule), and capital expenditure required to deliver the project.

In Era 1 and Era 2, a schedule (whether it takes the form of a simple bar chart à la Gantt or a complex CPM schedule from Era 2) merely tells us what we want to have happen. But the production system dictates what will happen. The question then is how best to understand and influence the behavior of the overall project production system and the implications of the subordinate production systems that it is composed of. The concrete production system versus the site grading production system versus the steel fabrication production system—what is each system doing? They all need to work together.

Production systems are governed by science of operations. The industry's failure to recognize this is a critical oversight and the answer to addressing how to fix the project performance problem. If we want to understand how a production system is behaving, we must apply operations science. To do so, a simple framework follows based on four verbs, five levers, and three curves. Or to make it easy to remember, the 4-5-3 framework.

THE FOUR VERBS

Design, make, transport, and *build.*

Those are what I call the four primary verbs. Most architecture, engineering, and construction operations fall into one of those four categories. Those verbs also provide part of the road map into Era 3 and a means of shifting the focus away from administration and toward production: how we design things (including the definition of requirements, concept design, engineering, detailed engineering, and production engineering); how we make things (manufacturing, fabrication, where we do what, how we do it); how we transport things (how we deliver it to site, the means of transport of materials, and how we prepare it for transport); and finally, how we build things (receive materials, information, and equipment at one or several sites upon which we are going to move the materials into place; erect/install/set/ place it; test it; commission it; and start it up).

By focusing on the specific tasks bounded by these four verbs (and in particular on imminent tasks), project production management shifts the focus from administrative work associated with project management as we know it today to the production elements of delivering a project. In the world of lean, it changes the focus from non-value-added to value-added activity.

This alone is a far-fetched concept to many. Whenever we ask a customer to send their process flow diagrams depicting their work processes, we almost always, to a tee, receive business process maps (i.e., showing the estimating, scheduling, and procurement processes). As stated in the LCI charter, the industry is blind to production.

FIVE LEVERS

Era 2 proposed an iron triangle of time/cost/quality trade-off—you can have two but not three. But in Era 3, we are no longer beholden to this trade-off. In contrast, I posit that there are five levers that affect cost, time, and use of cash. By using them, we can break the inhibitory grip of the iron triangle. These levers are product design, process design, capacity, inventory, and variability.

1. Design of the product: This pertains to everything from the design of the entire facility to its most granular details, such as how a pipe is to be connected to a valve. The product can be final assembly of the facility or just address the most

discrete parts. How we design those things has implications on the process we use to make and install them.

2. Process design: This lever exists in tandem with product design—how do we need to do the work to create that part or product, and how can we do it better? If we're going to weld two pipes together, that's different from connecting the pipes together with a coupling. The process is different, the equipment is different, and the required skill is different.

There's a relationship between product and process. In the US especially, because of various lawsuits and the stipulations of insurance policies, architectural and engineering firms are generally not allowed to get involved in means and methods—they cannot dictate the process even though the product they designed dictates that process. This is a source of great consternation in the industry because many things are designed on the product side with no input from the process side. And by the time the implications for process are understood, it's too late.

There's a relationship between product and process.

Later in the book, we'll talk about production engineering, which is common in the automotive sector and other industries but is underutilized in construction. We'll also address concurrent engineering (another hallmark of the auto industry), which means designing product and process simultaneously.

3. Capacity: The maximum average rate at which the items/units/tasks/products can flow through a process or system. Capacity is the upper limit of throughput and its contributors

(labor, equipment, and space). The product flows through the process and uses labor, equipment, and space to do the work. You have two pipes that need to be welded together; some guy (labor) with a welder (equipment) in a shop (space) is going to weld them together.

4. Inventory: There are three types of inventory.

   a. Inbound stock

   b. Work in process (WIP) includes things being worked on and things waiting to be worked on (queue)

   c. Outbound (finished goods)

   At the beginning you've got a stack of pipes (inbound); later, pipes being cut/cleaned/welded (how much stuff is being worked on or waiting to be worked on: WIP); and then pipes that go out to the site or the customer when finished. The pipe spools ready to be shipped are finished goods to the fabricator and when delivered to the site are inbound raw materials to the installer, but all of which is WIP: tied-up cash to the owner! A key concept is that inventory, especially WIP, is the proxy for time. How do you determine or predict time in a production system? Calculate the WIP.

5. Variability: This refers to, broadly speaking, anything that is different from something else; in particular, it is variability between time required for a given task at a specific point in the process. If the pipe can be cut in one minute, it takes one hour to weld it, and ten minutes to test it, and you keep cutting pipes, what happens? A backlog—a queue.

   Reasons for variability can be both beneficial and detrimental. Beneficial variability is a late change in the design process or even after construction begins. The change has a negative impact on the project but a beneficial impact to the operation of the asset and the value to the business. The cost of the change is far offset by the value to the business for making the change. An example of detrimental variability may be a late delivery which does not create value, or a faulty weld.

Taylor would try to speed up every worker along the line to maximize the output of each, but as we discussed before, that's counterproductive: due to variability pipes start stacking up at the slowest point in the process (the welder, in this case), which incurs a cost (not an abstract one; a real cost, in dollars). It's Era 1 thinking that dictates "cut as many pipes as you can regardless of how many you can weld or test."

The bottleneck is the welding, so we might as well operate at the level of the bottleneck. But all that is happening because of dependence and variability: the welder is dependent on the cutter. If there's any variability, either the welder must wait or the cut pipes stack up.

In construction, the goal is to never allow labor to wait on materials. It is so engrained in the current thinking that performance measurements focus on: If and for what duration does labor wait for materials? In response, construction professionals and academics develop methods based on using an inventory of materials to ensure labor never waits. Again, the focus must be on throughput, not productivity and utilization.

The iron triangle holds that the project deliverable and the process for delivering are immutable and inflexible. Era 3's emphasis on project production management negates that limitation by considering that the product being delivered and the process by which it is delivered might be designed differently. Along with capacity, inventory, and variability, we now have different levers that let us minimize cost and time while increasing scope. By applying operations science to compute the use of a resource, we can guarantee that it does not surpass a limit determined by the inherent variability. Other factors affecting the variability of the process include design specifications, logistics, technology, worker skill, and the environment where the work is being performed. But there is more: we must understand the important production-based relationships.

THREE CURVES

In the world of OS, there are three curves, derived from three mathematical equations, that represent the dynamics between cycle time, utilization, throughput, and WIP. Understanding the curves and applying the underlying equations provide the intuition and foundation for optimizing virtually any production system, from high-volume line flow manufacturing to low-volume fabrication and everything in between.

Curves

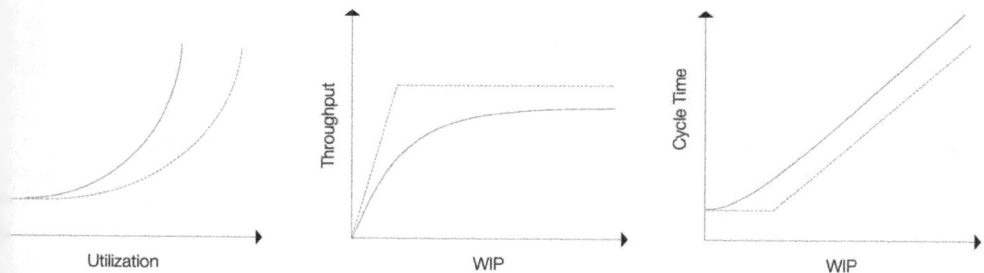

Three Fundamental Relationships of Operations Science

1. The relationship between cycle time and utilization/capacity. The higher the utilization, the longer it takes. On the freeway, the more cars you put on the freeway—in other words, the more the freeway is utilized—the longer it takes to get from point A to point B due to traffic.

2. The relationship between throughput (how much stuff gets through the system in a given amount of time) and its relationship to WIP (work in process). As the curve indicates, you can continue to "push work in" only so much: at a certain point, throughput plateaus. That is, the cutting station keeps cutting pipe, but the bottleneck of the welding (which takes more time per pipe) causes it to max out. Per the first curve, the more you more you introduce variability, at the same utilization, the more time is required (relationship of the first curve). Continuing with our freeway analogy, how do traffic engineers solve this problem? They install metering lights so that only so many cars use the capacity of the freeway (relationship of the third curve).

3. The relationship between WIP and cycle time. The higher the WIP, the longer it takes. Let's take a moment to think about our friends in Era 1 and Era 2. Taylor's goal was to drive utilization as high as possible. You may not have heard it from Taylor's mouth, but if you work in construction, you're doing what he said to do, even if you don't know it. Meanwhile, the goal of Era 2–type schedulers is to cram as much work into the production system as possible in an effort to burn and earn. But as curve number three (the correlation between cycle time and WIP) demonstrates, that just extends the life of the project.

Era 1 and Era 2 are flawed because they don't understand that the projects are an assemblage of production systems behaving in accordance with the parameters of operations science. Again, that's the industry's giant blind spot. They don't understand the OS math—represented by these curves—that undergirds everything we do as we design, make, and build, whether on-site or off-site, by hand or with robots.

I know that industry professionals are oblivious to this fact because I've seen it with my own eyes, not just on the jobsite but in seminars and conferences around the world, where the smartest people in the industry congregate. When my colleagues or I present at such gatherings, we often do an exercise that is edifying for everyone in the room. The question is simple: we ask the attendees to describe, with a quick sketch, what the relationships are between cycle time, utilization, throughput, and so forth. This is Operations Science 101, but most people have no idea. If there are twenty people in the room, we get numerous different answers.

This is illustrated, literally, when we ask them to come up and draw each curve on a chart. The x and y axes are already written; all they have to do is approximate the curve. But it's really not their fault, and in reality, it is not funny. It is perhaps the single reason the construction industry struggles to deliver projects in accordance with objectives. In any industry, people only know what they're taught, and there's no fundamental understanding of OS in the world of construction because PMI states that operations management (OM) and its overarching field, operations management, are outside of project management. It's something you apply after the project is complete, not, as I contend, a body of knowledge used to manage projects better.

OS is the study of the relationships represented by these curves. The field boils down to a series of mathematical equations that allow you to understand, predict, and influence what happens in production. For those who would like to understand these equations, they are shown below.

Little's Law:

CT = WIP / TH

Cycle Time Formula:

CT = RPT + BT + MT + QT + SDT + WTMT + PTB

RPT = PT + ST + DT
BT = (Waiting for Batch) + (Waiting in Batch)

VUT Equation:
$$CT_q \approx V \times U \times t$$

$$\approx \left(\frac{c_a^2 + c_e^2}{2} \right) \left(\frac{u}{1-u} \right) t_e$$

Copyright Factory Physics, Inc.

Let's revisit PMI's definition of a project, as described in the PMBOK: "a temporary endeavor undertaken to create a unique product, service, or result."[12] Later, the guide states that "operations management is a subject area that is outside the scope of formal project management as described in this standard. Operations management is an area of management concerned with ongoing production of goods and/or services. It involves ensuring that business operations continue efficiently by using the optimum resources needed and meeting consumer demands. It is concerned with managing processes that transform inputs ... into outputs."[13]

---

12   Project Management Institute, *A Guide to the Project Management Body of Knowledge (PMBOK Guide)*, 2012, fifth ed., Kindle.

13   Project Management Institute, *PMBOK Guide*, Kindle locations 601–602.

PMI's repudiation of OM as relevant to construction turns on its erroneous understanding of the word *ongoing*. Are construction projects one-off endeavors, or continual sites of production that involve a bevy of repetitive processes? It's hard to make the case for the former. Remember the four verbs: these operations occur in sequence, sometimes hundreds or thousands of times over, and for a period of weeks, months, and years. Furthermore, as stated earlier in the chapter, the majority of production systems that form the overall project production systems are most often serving other customers on an ongoing basis. That is the very definition of *ongoing*. Therefore, the kind of flow methods that are applied, to great effect, in manufacturing are likewise applicable to construction—despite PMI's insistence to the contrary.

Not only does OS enable us to understand and influence a production system's behavior, but OS can also be used to understand the various strategies and methods being promoted in the market. This includes lean construction, advanced work packaging, workface planning, etc. By understanding the science, we can decipher what is being proposed and what will most likely happen when the strategy is implemented.

Lean construction is a frequently used buzzword, but most consultants and organizations who tout their belief in lean construction are mostly just spouting off rhetoric, without a clear means of applying lean construction principles. Granted, the lean methods that work well for Toyota are not easily transferable to a complex construction project. But that's not because it's not feasible. If construction professionals have reduced lean construction to jargon or a buzzword, it's because they don't understand that operations science is the key that unlocks the door.

The promise of advanced work packaging is that in exchange for more investment in planning and by using stocks of stuff, we can increase the amount of "time on tools" (time workers are working). Yes, inventory buffers of predecessor work and materials will most often increase time on tools and allow a task to get executed faster. But this is tied to the stubborn belief in Taylorism: the more they work, the more that comes out. That notion is fallacious because it overlooks the bottleneck effect and basic production system dynamics, including the three curves.

The industry ignores the bottleneck because of an excessive preoccupation with making sure all workers are never waiting on materials. That's a cardinal sin in the industry! It is not uncommon for aircraft to be chartered on a routine basis to ensure materials get to a site.

Back in the contracting days, we were doing a project in China at a new semiconductor plant. One element of our scope was installation of the roof assembly. One day we received an email stating that the plant that made the insulation for the project had a problem and that they needed to replace the product. The manufacturer was horrified to learn that the material had landed in China but there was not much that could be done as they had already put us on notice regarding the defect. But what was to come was even more horrifying for the manufacturer. The CM firm said something to the effect of, "No problem, but to ensure we do not suffer any schedule delays, we have chartered a JAL 747 supercargo plane to collect the replacement product from the West Coast and transport it to the site in China next week. To make the paperwork easy, we will just back charge your account." Needless to say, the cost of the air transport was more than the cost of the insulation.

But it's fine and even preferred if materials are waiting on workers, so there's no aversion to letting inventory stack up at various points in the production system. This, too, is problematic for a number of reasons I'll get into in the chapter to come. But basically, WIP buildup and inventory incur costs (labor equipment space, use of cash) to amass it, handle it, hold it, and preserve it. There can be theft (not necessarily criminals, though some sites experience this, but craft people removing one part to use somewhere else) and damage due to excessive handling or lack of proper protection. Obsolescence is another major factor as sometimes equipment warranties expire. Sometimes the design changes make that $50 million spent on materials result in those materials no longer being needed.

Increasing the amount of WIP just spikes the amount of time it takes, as illustrated in the curve showing the relationship between cycle time and WIP. But in Era 2, earned value management encourages putting more work in. So that concept results in the unintended consequence of the projects costing more, taking longer, and tying up more resources. The following graph from an actual project depicts the relationship between WIP and schedule duration.

Project Duration vs WIP

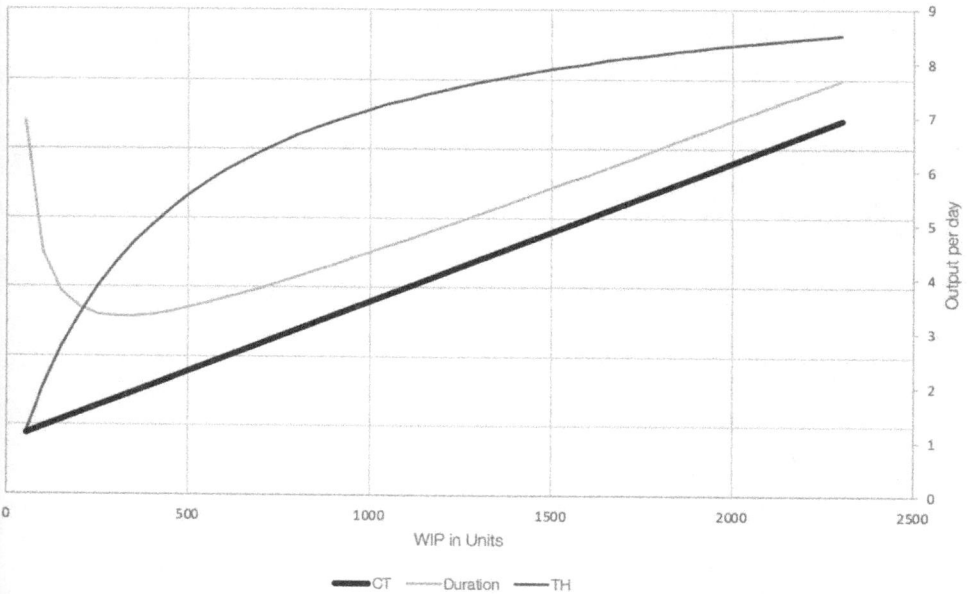

Relationship Between WIP and Project Schedule Duration

If there is not enough WIP, the project is delayed for lack of throughput, but if there is too much WIP, the project will also be delayed, as the capacity of the system is reduced due to congestion and the work takes longer to complete.

The following project controls progress graph shows planned versus actual for the fabrication and assembly of a jacket and topside for an offshore platform. The graph depicts the project being on schedule for a few months, then trending in the wrong direction. There was going to be more work than there was time and money. But equally important, the typical project controls report does not indicate what is wrong other than that they are not on plan. It does not say what to go do about it.

We have the *what* (we are late); we have the *so what* (every day will cost millions in lost revenue), but what we don't know is what to do about it. In this project, to optimize their production, the contractor rolled all the mill elements but did not begin the assembly process until all elements were rolled (many still needed the seams to be welded).

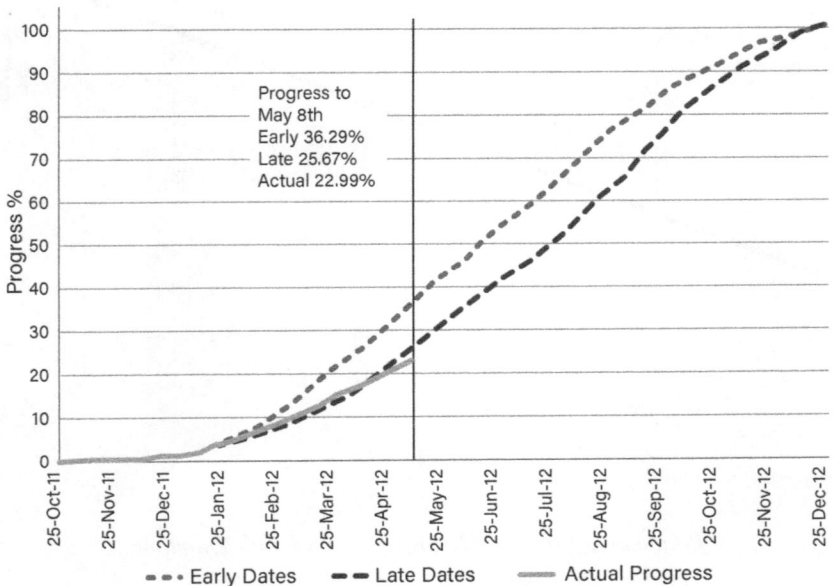

Progress to
May 8th
Early 36.29%
Late 25.67%
Actual 22.99%

- - - Early Dates   - - Late Dates   —— Actual Progress

On one very large project in Australia, the company contracted a cruise ship to hold additional craft workers. Eventually, the ship was removed, and with it, its many craft workers. What happened next? More work got done. How can that be?

Because the surplus of workers was just creating increased WIP. Stuffing a project full of workers cannot invalidate the law of the bottleneck no matter how many cruise ships you have docked off the Gold Coast. The bottleneck is the bottleneck. You can cut pipes all day and night, but if you can't weld them at the same rate, there's a queue. Cost and risk increase as time gets extended. Ironically, most often labor cost continues to increase as workflow gets out of sequence, but there is no shortage of construction management firms calling for more people. Back in the contracting days, we had a customer so eager to get more people, we parked a truck with our logo next to his office. He saw the truck, figured we sent more people, and was happy!

To truly understand the principles and equations of operations science, we propose you read *Factory Physics*, written by Mark Spearman and Wallace Hopp, absolutely two of the most brilliant men who have ever been involved in the field of operations and supply chain.

THREE TYPES OF CAPITAL PROJECTS

Adding to the 4-5-3 framework, we can establish there are three fundamental types of projects with differing challenges and management requirements.

1. **Construction**: of a building or a plant. Materials are delivered to one general site or place. All production activity is concentrated there or in shops in support of the site.

2. **Deployment**: examples include a communications network, cell towers, or renewable energy technology that needs to be installed at numerous locations. Instead of production being concentrated on one site, we're delivering information, materials, equipment, and perhaps people to multiple or many sites. Therein lies the challenge: not everything is going to a single place. Moreover, this type of capital project requires interfacing with numerous municipalities and jurisdictions to secure a bevy of permits and permissions and technical design and engineering approvals. Can we dig up this street? Can we place an antenna on that building? Et cetera.

3. **Maintenance/repair/overhaul**: in these projects, we have to first do work to figure out what work we have to do. For example, there's an existing facility, but we don't fully understand what's behind the wallboard or enclosed in some piece of equipment. We need to take stuff apart to ascertain what we're working with.

Each type requires a different approach as to how resources are planned and used. For instance, a construction project requires ongoing management of supply to a single site and coordination of work at a single site. Materials should be delivered on an as-needed basis, wherein a maintenance project that by nature results in more variability than a construction project will benefit from having extra capacity and inventory of raw materials and parts to make rapidly on demand and deliver fast. This is just like the difference between a consumer oil-change shop doing an oil change for your personal car and a pit stop for a Formula 1 racing team. The oil-change shop and you are willing to wait a bit, whereas in an F1, races can be won or lost in the pits. From an OS perspective, the oil-change shop will let time get extended to optimize use of capacity (mechanic, shop, and tools), as opposed to an F1 team that has numerous people, tires, and other parts ready to go. F1 teams are willing to pay the cost associated with having excess capacity and inventory—but if an oil-change shop did the same and didn't have the demand, they would go broke.

PROJECT PRODUCTION MANAGEMENT

The application of operation sciences to the delivery of a capital project is done using specific methodologies and enabling technologies under the umbrella of project production management, or PPM. These methodologies include **production system modeling and optimization (PSO)**, **project production control (PPC)**, including **computer-aided production engineering (CAPE)** and **supply flow control (SFC)**. It is important to note that PPM is not meant to replace all the collaborative ways of working offered by industry bodies and consultants but rather to provide the underlying technical framework for thinking about how to deliver a capital project in accordance with the project's purpose and objectives.

The application of these PPM methods directly addresses the gap

> The application of operation sciences to the delivery of a capital project is done using specific methodologies and enabling technologies.

explained earlier in the book. The use of these methods has generated billions of dollars in value for owners and contractors undertaking complex and critical capital projects, from civil infrastructure to large-scale upstream energy projects onshore and offshore, as well as commercial buildings.

Though PPM supports and even incorporates a continuous improvement element including the use of Shewhart's plan-do-study-act methodology (some credit Deming, who made it more well known through the Deming cycle), PPM is based on a "define, design, and control" approach. Business objectives are identified and defined, the production system is designed or optimized, and then the production system is controlled. Continuous improvement is a subset of control. Another way to look at it is that the define-and-design/optimize element is strategic and overarching, where continual improvement is incremental or tactical.

The PPM framework provides the means to identify and remove the use of unnecessary resources contained in a project production system, whether early in the project-definition phase or when a team has a project that goes sideways, or as Roberto Arbulu of SPS says, becomes "a big fucking problem." From a deployment perspective, PSO may precede PPC or SFC, or PPC may precede PSO. When deploying PPC, standard work should also be developed and leverage CAPE when appropriate.

With one recent customer, we started with SFC then moved to PPC and will look to use PSO last. In other words, PPM and its three primary methods support design of new production systems and improvement of existing production systems. Kind of like fire prevention or teeth cleaning versus firefighting and root canals. Some do prefer the excitement of the danger and pain!

The construction industry prides itself on its commitment to environmental health and safety, but in an industry that focuses on administration over production, we must question how serious we are. A world-class manufacturing company asked us to participate in their annual safety conference. They would go around the table and ask each expert what they recommended. Our response was always the same: eliminate any unnecessary work and associated use of resources. One year one of the leaders said, "Why don't you come to my facility and show me and my team what you are talking about?" Of course we took him up on the opportunity, and when we left, they went to work (next year they proudly presented what they had accomplished).

The following section outlines a technical approach for doing just that—identify and eliminate unnecessary work and associated use of resources, all done in accordance with project and business objectives.

As outlined in the following diagram, to ensure that a production system behaves as intended (conforms to desired objectives), two key elements are needed: (1) understanding and influencing how the production system will behave and (2) effective control of the production system.

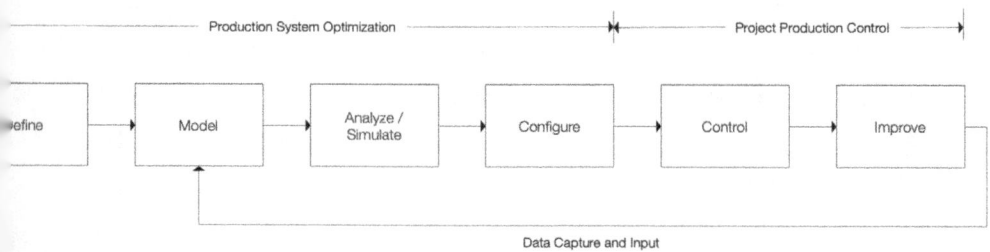

Relationship between Production System Optimization and
Project Production Control

PRODUCTION SYSTEM MODELING AND OPTIMIZATION

When it comes to diagnosing a problem or making something better, there seem to be two options: (1) tinker with it and (2) follow a framework. Optimizing a production system should be no different from visiting a medical doctor where you expect the doctor, through questioning and various tests, to diagnose the issue. You don't expect them to tinker. To do this, the doctor must have an underlying framework and must be looking for something. At SPS we always endeavor to look *for* it (i.e., Where is the WIP building up? Where is the capacity being lost? Where is there excess capacity? Et cetera) rather than look *at* it. Looking at it becomes complex and overwhelming. A PSO study is no different—you must start with a framework, and operations science provides the necessary framework. That said, many construction professionals are willing to tinker.

To describe it simply, inbound items flow through the production system and become finished goods. Items can only flow through the process as fast as the bottleneck will allow. The items that have entered the production system and have not been made into finished goods are WIP. If we want to get more through the system, while minimizing cost and use of cash, we need to optimize the capacity and the WIP while mitigating any detrimental variability. A key element is the flow through the bottleneck. In a high-performance production system, the bottleneck is purposely located as sort of a control point. It acts like a governor or regulator. This is in contrast to lean consultants looking to "balance" all operations or process centers in the production system.

PSO is the practice of mapping, modeling, analyzing, and simulating a production system to achieve the best-desired performance. PSO studies can be used to improve any and all types of production systems and supply chains. Key areas of focus for PSO are throughput, cycle time, and use of resources (capacity and inventory), along with sources and implications of variability.

Much like how engineers use models to understand the anticipated behavior of a mechanical or structural system, PSO models can be used to understand the behavior of a production system. Unlike project management that centers on dates for when work needs to be complete, production models focus on key production performance indicators that ultimately drive schedule dates including throughput, cycle time, and capacity utilization, to name a few. As stated in the previous chapter, other than product and process design, the only remaining levers to improve production system performance are (1) regulating capacity utilization, (2) control of WIP, and (3) management of variability.

A PSO study begins with identification and definition of a production system that will benefit from a PSO initiative. The subject production system may be technical work carried out during design, fabrication, and assembly work being performed in a shop; construction work being executed on a site; a flow of supply; or some combination thereof. In construction, identification and definition of a production system is most often accomplished through the creation of a production system map. Mapping and modeling of a production system is based on using a specific set of symbols that allow readers to quickly understand what is happening in each step or operation in the process. Barrels are used to denote stocks, triangles to depict queues, and rectangles to represent operations/process centers. Arrows are used to indicate flows or routings.

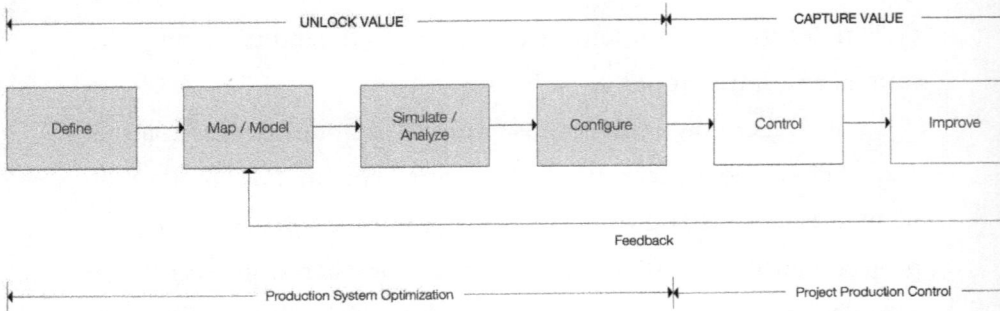

I remember teaching a class on how to do PSO at Stanford with James Choo of SPS. Several students were attempting to use the PSO application to create bar charts. The best part was when a student would select the "organize process" button and the bar chart would reorganize into a process flow diagram—and the student would jump back a bit from the monitor, then say there is something wrong with the scheduling software. We watched in amazement and, I must admit, found it a bit entertaining. It is astounding how deeply rooted this thinking has become and to see the value destruction it causes.

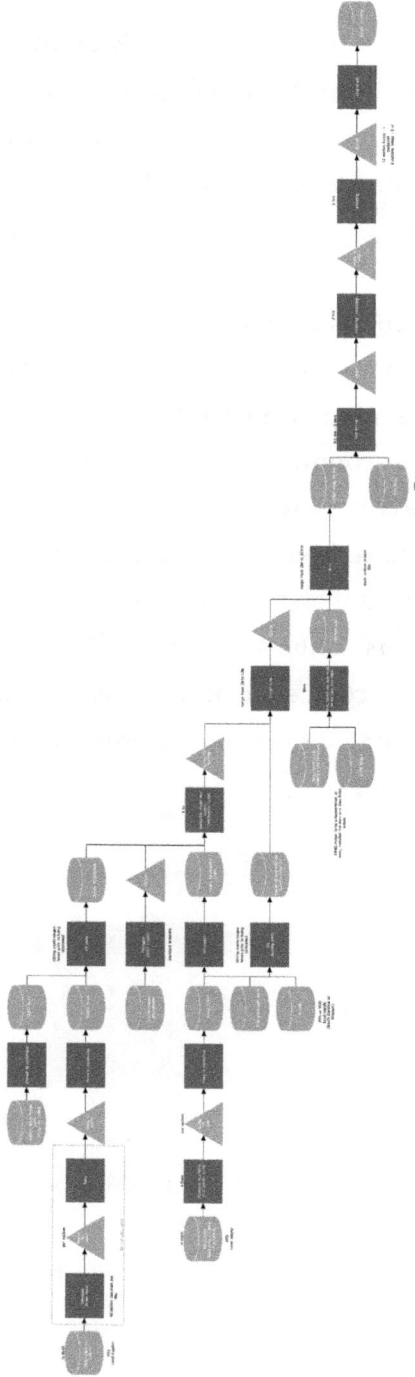

Project Production System Process Flow Diagram

After the production system is identified and defined, the next step is to transform the map (which is dumb) into a model with intelligence. This is done through adding production system performance data such as desired, predicted, or actual throughput, cycle time, capacity utilization, and a variability factor, among other production-based parameters. Cost can also be included to enable financial analysis.

Once the map becomes a model, various modeling methods are used, with analytics and discrete event simulation being the most common. Often both are used in tandem. When the production system uses a production control application in the case of an assembly line, the data from the control system can be entered directly into the PSO application. As stated later in the book, the capture of this data can now be automated through the integration of production control systems and IoT sensors connected to the PSO model, creating a digital twin of the production system. This sets the stage for what is to come, the self-forming, self-optimizing production system (more on that later in the book).

From "Dumb" Process Flow Diagram to Intelligent Production System Model

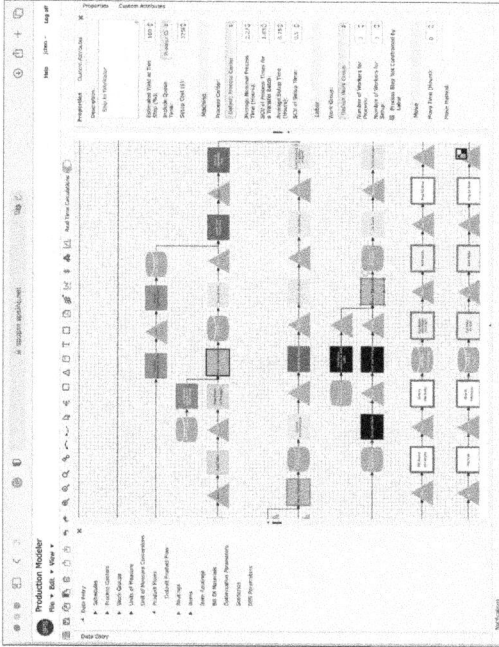

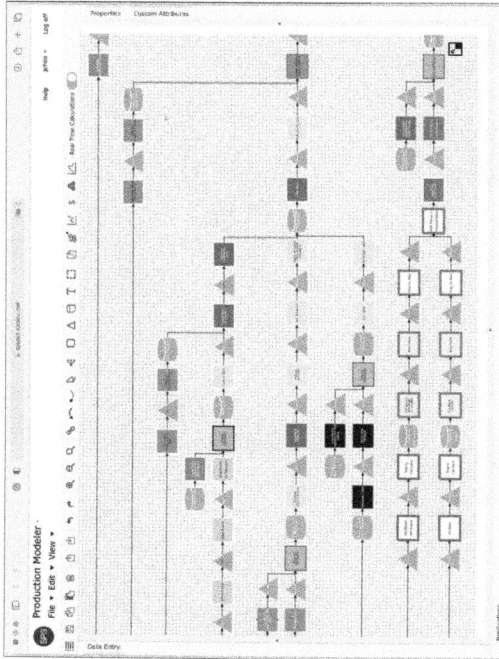

The model is then validated (ensuring the model behaves as the production system does or will). The validated model is then used to analyze, simulate, and optimize various production parameters, including WIP and use of capacity, in accordance with desired objectives.

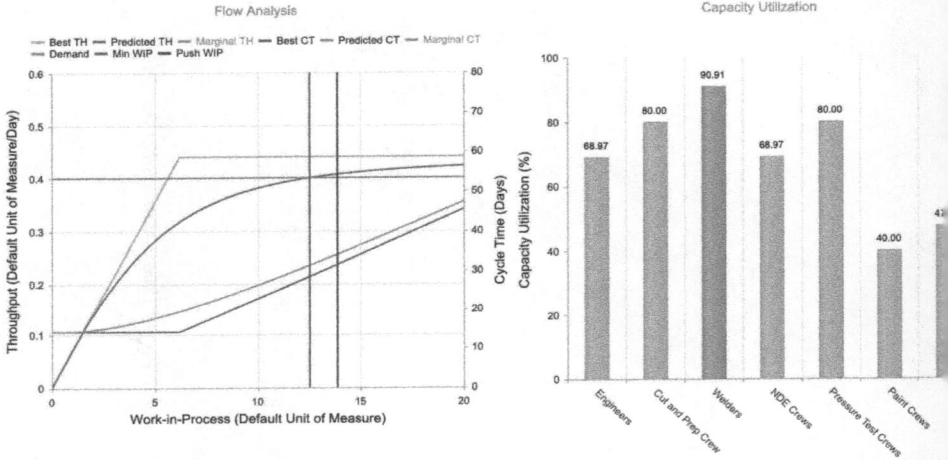

Output from Analytical Model

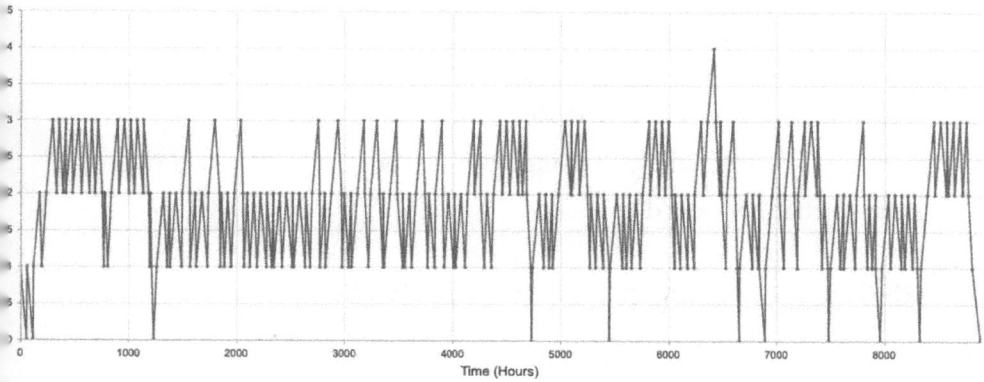

Item	Cycle Time (Hours)				Process Time (Hours)				Queue Time (Hours)			
	Avg	SD	Min	Max	Avg	SD	Min	Max	Avg	SD	Min	Max
	110.95	37.73	27.77	240.31	106.34	36.64	27.17	240.31	4.61	12.33	0.00	80.68

›cess Center	WIP in Queue (Unit of Production)			
	Avg	SD	Min	Max
d Prep Crew	0.50	0.50	0.00	1.00
s	0.50	0.50	0.00	1.00

rategic Project Solutions, Inc.

Discrete Event Simulation Model

Following are examples that show how PSO can optimize a production system. The first was done during the design phase, where we worked with McKinsey, whom we often partner with. The second was done with a fabrication shop. In this example, we can see how the throughput can be increased at the same time the cycle time is decreased. Financially, this equates to enhanced customer service and more cash in with less cash burn. The third is for the installation of cabling at a construction site where time can be compressed by reducing capacity and WIP. The fourth, another project done with McKinsey, shows the potential optimization for two discrete civil infrastructure projects.

PSO for Engineering Work

This project supported the design and deployment of a mission-critical communications network (as in national defense). In this case, it was imperative to ensure design would be delivered complete and on time. The PSO analysis along with various other remedial efforts ensured that happened.

PSO to Optimize a Fabrication Shop

Here we can see that throughput can be increased while cycle time is reduced by optimizing capacity and controlling WIP. Do more, faster, for less!

PSO to Optimize Construction Site Operations

In the next examples, we can see that the work can be completed earlier using less capacity and less WIP. What needs to happen is that the capacity strategy needs to be rethought, including reducing the unnecessary staffing.

Comparable cycle time (normalized), in minutes

Legend: ■ Decouple from current process ■ Improving work conditions ■ Further improvement with PPM

	Actual, 2 piles / day	Improvement "standard" lean, 3 piles / day	Further improvement with PPM, 5-6 piles / day
Sling installation	5	1	0
Lifting and positioning bottom pile	25	25	0
Sling removal	5	1	0
Position pile-driver	5	5	0
Pile driving bottom pile	1	1	0
Pile-driver park	3	3	0
Lifting and positioning top pile	2	2	0
Guide welding	10	5	0
Positioning and lowering top pile	5	5	0
Pile joint welding	90	60	60
Welder rest & walk to next joint	0	0	15
Sling removal	5	1	0
Position pile-driver	5	5	0
Pile-driving top pile	10	10	0
Lunch break (prorated)	30	20	10
Non-value adding time (prorated)	60	40	20
Total	261	184	105
Real cycle time	171	124	75
Shift time	9 hours (8am-5pm)	9 hours (8am-5pm)	11 hours (8am-7pm)

Callouts:

(1) Add 2nd set of slings to decouple from crane

(2)(3)(1) Reduce bottlenecks through additional welders and improved work area

(4) Add 2nd crane[1]

Setting and positioning of the middle pile done by the other crane

Add 2 working hours

Welders as key driver are idle ~50% of time
Bottleneck: crane

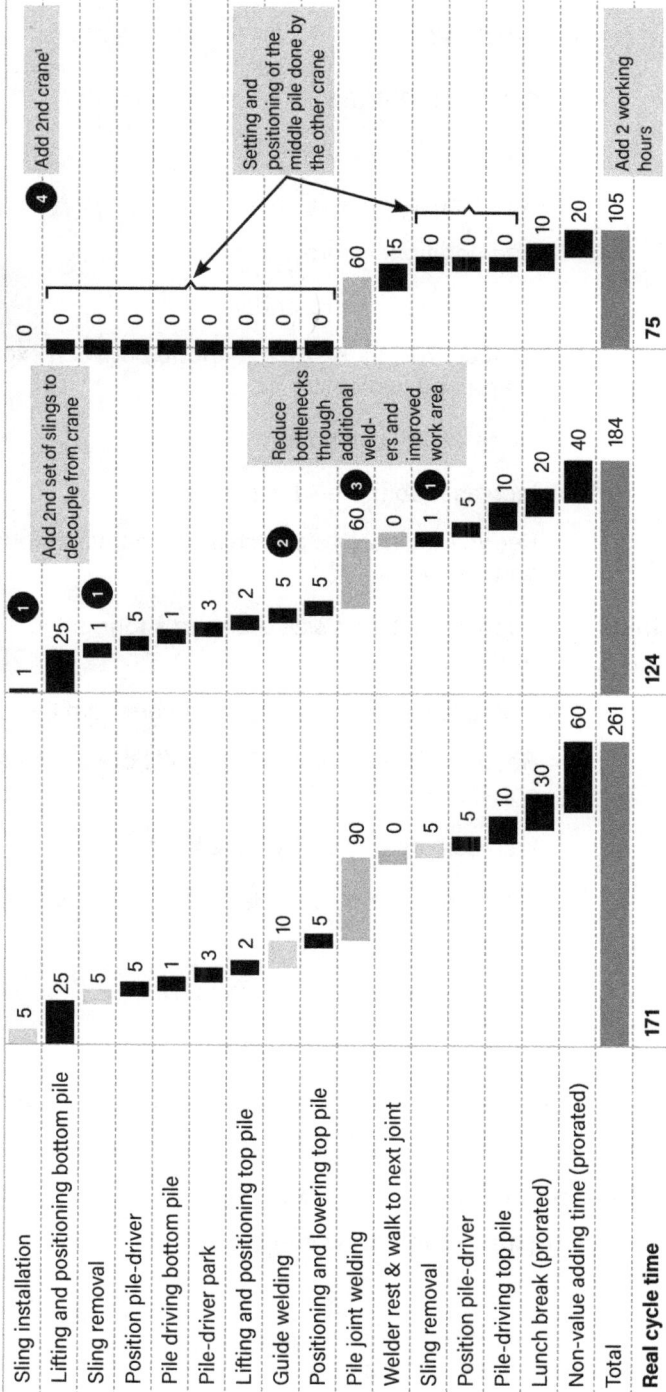

Welders are close to fully utilized
Bottleneck: welder

1. Low feasibility – constricted space due to the presence of high voltage line

Source: McKinsey & Company

89

SUMMARY OF OPPORTUNITIES

WORKFLOW	OPPORTUNITIES	IMPACT
Precast Piling	Add additional crane	11.5 days
	Add time to work calendar	
Steel Sheet Piling	Add additional vibro-excavator	10 days
	Add time to work calendar	
Pile Cap Construction	Reduce the batch size of handoff between excavation and pile cutoff	22 days
	Extend work calendar	
	Increase rebar crew by 2 men	
Column Construction	Reduce the batch size of handoff between rebar and formwork	8 days
	Reduce the batch size between shoring and rebar	
	Increase the # of formwork workers from 8 to 16	
Pier Head Construction	Reduce the batch size of handoff between rebar and formwork	13 days
	Increase rebar crew by 2 men	
	Add time to work calendar	

Source: McKinsey & Company

Since a production system model is just that, a model, it is imperative that the production system's behavior be controlled in accordance with the production system objectives. This is the purpose of production control and supply flow control.

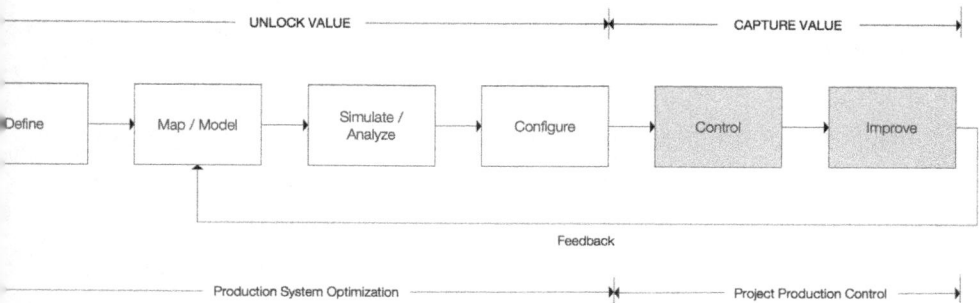

PRODUCTION CONTROL

An effective means of production control is a key element of any efficient and reliable production system. Technically, production control is any action, process, mechanism, system, or combination that organizes and enables control of production, or work execution. Production control uses physical, software, and human decision-making for the control-to-control production routings/sequence of work, use of capacity, and the amount of inventory, including WIP. By *physical* we mean actual devices that control what happens, much like the plate that forms the size of the entrance to a baggage scanning machine at the airport, or the number of cranes on a construction site. *Control* means actual allocation of capacity, management of minimum and maximum levels of WIP and management of variability.

It is important to understand the various control protocols, schedule (known as *push*), pull, and constant work in process, or CONWIP, including how each works. Push systems use predetermined schedule dates to release work into the production system. Though most widely used in construction, push, or working to a predetermined schedule, is not effective means of control due to the impact of variability. Developed by Toyota, the *pull* control protocol is

91

based on the downstream operation sending a notice—or, in the case of Toyota, a kanban card—to the upstream operation to produce. The CONWIP control protocol sends a signal to the beginning or near beginning of a production process and, in so doing, controls the WIP between the operation that initiated the signal and to the operation where the signal was sent.

Lean experts go on about push versus pull, but as Mark Spearman says, "Push, pull, what's the point. The objective is to control the WIP."

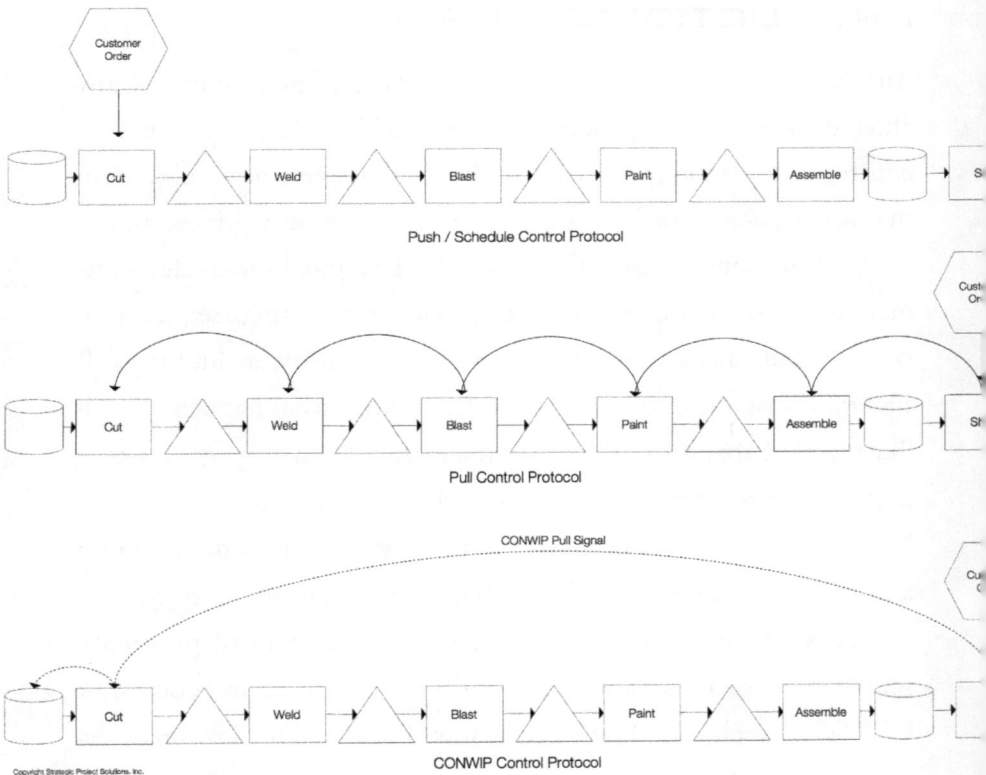

Push / Schedule Control Protocol

Pull Control Protocol

CONWIP Control Protocol

From a design perspective, the strategy should be to use a pull or CONWIP control protocol whenever possible. Push systems, including the use of schedules, should be used as a last resort.

PROJECT PRODUCTION CONTROL

Production control software has been used in manufacturing for decades, while construction is overreliant on human decision-making as a means of control. The advent of project production control (PPC) now provides the software element of the production control system for construction projects.

PPC is used to effectively control production for both technical work (design and engineering) and physical work, whether in a fabrication facility or on a construction site. An effective PPC system must ensure that (1) the sequence of the work is followed, (2) resources including capacity are being allocated efficiently and WIP maintained within the minimum and maximum levels, and (3) that variability and its sources are managed.

Though project controls and project production control may share some of the same business processes and data presentation (e.g., a network diagram or bar chart) PPC is not "scheduling" in the vein of Era 2. It's something altogether different—and better.

Since the goal of production control is to manage the sequence of work, use of resources, and variability, including from its sources, CPM scheduling is not well suited as the basis for production control. Additionally, and very important to understand, production control is not the same as taking a bar chart or CPM schedule to a greater level of detail (i.e., from weekly to daily).

We are not talking controls as in forecasting and reporting. We are also not talking scheduling even if you ask the players to do a collaborative pull plan. Peter Drucker described the differences between controls and control as follows: "The word 'controls' is not the plural of the word 'control' ... the two words have different meanings altogether. The synonyms for controls are 'measurements' and 'information.' The synonym for control is direction ... Controls deal with facts, that is with events of the past. Control deals with expectations, that is, with the future."[14]

Because production rates drive schedule dates, typical project controls reports, such as cost and schedule performance as well as cash flow analysis, become an automated output. As you will read in chapter 10, the use of current technology enables all this to be integrated and automated, providing real-time scheduling and cost forecasting/reporting along with the use of AI and ML (machine learning) by companies such as Chevron.

LAST RESPONSIBLE MOMENT SCHEDULING

At SPS, we believe in, and we have developed and applied, the concept of last responsible moment (LRM) as the means for creating schedules. The LRM scheduling method enables management of WIP while ensuring capacity is invested efficiently.

Simply put, LRM scheduling is based on the identification of the LRM finish date (not to be confused with the "last *possible* moment" date) and then subtracting the cycle time for the end-to-end process to calculate the start date for the process.

Start date = last responsible moment finish date –
end-to-end process cycle time

---

14   Peter Drucker, Management: Tasks, Responsibilities, Practices, (New York: Harper & Row, 1974).

Because the goal is to control WIP, which has an inverse relationship with time, we aim to work until the last responsible moment. We only want to do work when we need to do work. We're essentially adopting a just-in-time approach to executing work on-site where we are doing things in sequence, as needed, to maximize the throughput of work—getting it complete in the sequence we want rather than just doing work to burn hours and spike our burn and earn.

Software-based PPC solutions do this through three primary business processes: production scheduling, production planning, and, as stated above, continuous improvement. Production schedules are created using standard work processes to the extent possible and look out beyond the production planning cycle, which may be by the shift, the day, or the week depending on what type of work is being controlled.

> **Because the goal is to control WIP, which has an inverse relationship with time, we aim to work until the last responsible moment.**

Production schedules provide the forecast of what needs to be completed when and by whom. Production schedule dates are calculated using the LRM date. Production plans focus on the allocation of work for a specific period and provide the basis for committing resources. These plans are the means for liquidating work in a safe and efficient manner while production schedules ensure that sequence and the associated WIP are effectively managed.

Example

In 2008, we were asked to get involved with a refinery project in the Midwest. The owner reported they had more work than time and money. The first meeting was most interesting, showing what can happen when the stakes are high, as a fistfight between a large Texan and a thin midwesterner almost ensued over how to solve the problem. In the end, the owner decided to deploy production control in its simplest form.

First off, looking at the progress curve on the opposite page, we can see that the project planned to use more labor and take longer to complete. But using PPC to control use of capacity and labor, a once-challenging project was completed under budget and ahead of schedule. The ability to manage WIP and the results were amazing. So much so that the owner benchmarked this project against other projects at their other refineries and began to understand how powerful PPC can be.

METRIC	PROJECT W/PPC	PROJECT 2	PROJECT 3	PROJECT
PIPING	2.5 mhr/lf (~4 Avg Dia In)	4.0 mhr/lf (~4 Avg Dia In)	2.9 mhr/lf (~6 Avg Dia In)	3.01 mhr/l (? Avg Dia I
STEEL	20.9 mhr/ton	30 mhr/ton	26.7 mhr/ton	46.5 mhr/to
CONCRETE	9.2 mhr/cyd	11 mhr/cyd	10.5 mhr/cyd	24.8 mhr/cy
INSULATION	.44 mhr/sqft .24 mhr/lf		.31 mhr/sqft .45 mhr/lf	

Benchmarking of Troubled Project with Other Projects
at the Owner's Other Refineries

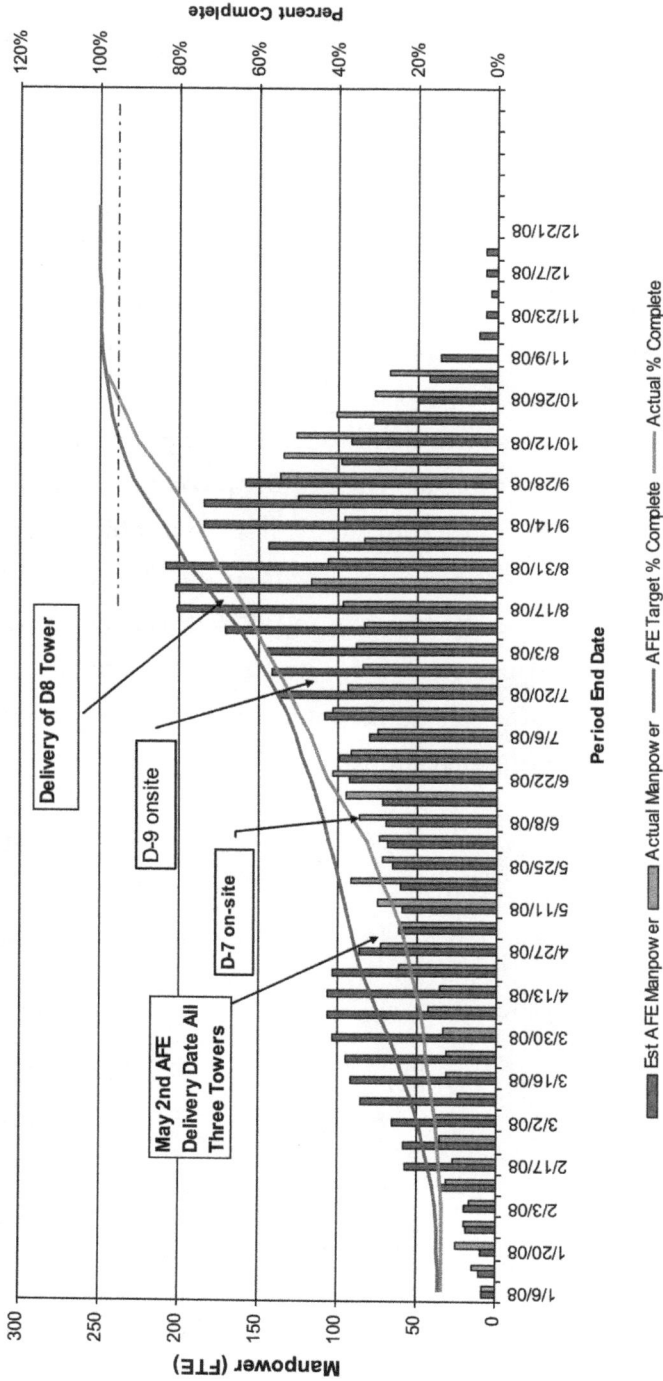

Manpower / Percent Complete

Outcome for Troubled Refinery Project

Example

Similar results were achieved at a large, heavy civil project in Australia where, due to various issues, the project was at risk of not achieving important milestones dates, including start-up and handover. The project controls team reported that all short- and long-term milestones were at risk. While talking on the phone with the project manager, I found he was more descriptive: "We are late getting later, over budget with the cost position continuing to erode, and the criticality is increasing as the press are all over us." It had turned into a real shit show, but he was committed to resolve the problem, and he did!

The deployment of PPC enabled the project team to gain time and ultimately deliver in accordance with schedule objectives and cost objectives. How was this accomplished? As stated, the implementation of PPC enabled effective control of WIP (including sequence of work), efficient use of capacity, and reduced variability. Use of standard work, LRM scheduling, and continuous improvement to optimize production. That simple!

Milestone Event	Planned Date	Actual Date	Variance
SP-1 Practical Completion	03 Aug 2012	19 July 12	-15 days
SP-1 Project Handover	12 Sep 2012	28 Aug 12	-15 days
SP-2 Practical Completion	12 Nov 12	25 Oct 12	-18 days
SP-2 Project Handover	22 Dec 12	04 Dec 12	-18 days

Courtesy of Strategic Project Solutions

STANDARD WORK

A key element of project production control is the use of standard work. The concept of standard work is not exactly groundbreaking; it's well understood and effectively applied in the automotive industry. They even have kinesiologists involved who ensure that the physical work a worker performs doesn't impose risk on a human body—and, of course, of deep interest to Taylor and the Gilbreths (their one best way), while Hauer went to great lengths to explain proper techniques associated with using a hand shovel, including correct placement of the feet and how to hold the shovel.

Surprisingly, many project professionals don't recognize that though the product, the final asset, might be custom, the *process* is mostly repetitive—whether it's in the world of design and engineering and it's technical in nature, or doing physical work (such as the trades). Think about it. You couldn't get a degree in architecture or engineering if we couldn't standardize, document, educate, and train people in those work processes. Likewise with craft work, if we couldn't standardize, document, educate, or train people, there wouldn't be trades (welders, pipe fitters, etc.). If we go down a couple of levels in the hierarchy, there is basically an endless amount of standard work that can be identified, mapped, optimized, controlled, and improved.

As stated above, standard work used to be common in construction and dates back to the time of Hauer and Taylor. It wasn't unusual for general contractors to have extensive libraries of standard work dictating how the work is to be done on the site. Era 2 and its focus on project management and construction management has undermined that paradigm of standard work. Years ago, I asked a field engineer at a construction management firm about a row of dusty binders he had up on his office shelf. "That?" he said, "That's standard work." "Do you use them?" I asked. "We haven't used them in years. We're not involved in the work. That's the subcontractors' problem! But I keep them in the remote possibility we self-perform again," he said.

While standard work is not a new idea, digital technology provides novel avenues for us to bring it back within the realm of standard practice and in so doing unlock and capture huge sums of value. The following graphic is from the standard process library associated with the above civil project in Australia. These processes contain detailed production information and enable rapid development of production schedules. Standard processes also form the basis for continual improvement.

#	Task	Date
2	Penstock / Stoplog Installation	13 Apr 11
3	Stoplogs	13 Apr 11
4	Rectification of Secondary Concrete Inlets	13 Apr 11
5	Bandscreen	13 Apr 11
6	Common Drains	13 Apr 11
7	Fit out of DN1500 Valving to Risers	13 Apr 11
8	Sump Pump Piping Header	13 Apr 11
9	Chemical Building Room Completion	14 Apr 11
12	Wash Water Treatment Bldg Room	15 Apr 11
13	Instrumention Installation for WW Bldg	15 Apr 11
15	Commissioning Chem Building	18 Apr 11
17	I&C - Magflow QA	18 Apr 11
32	Install Pressure Switch (SDS)	19 Apr 11
33	Install Pressure Indicator (SDS)	19 Apr 11
34	Install Flow Switch (SDS)	19 Apr 11
36	Install AIT (PVC)	19 Apr 11
37	Install CIT (SDS)	19 Apr 11
38	Install CIT (PVC)	19 Apr 11
40	Instrument Rack no Tundish	26 Apr 11
46	Install LCS	28 Apr 11
47	Install Cable Management	28 Apr 11
48	Install Pump (including HU sensor)	28 Apr 11
49	Install Pump (including temp signal)	28 Apr 11
51	Install ESD Relay Box	28 Apr 11
52	Install E-Stop	29 Apr 11
53	Install HMI	29 Apr 11
54	Install UPS (including PLC connection)	29 Apr 11
55	Install Supply to DB	29 Apr 11
56	Install supply to PLC	29 Apr 11
57	Install Pump (including HU and interlock)	29 Apr 11
58	Install LCS (with E-stop Relay connection)	29 Apr 11
59	Install Polipak	29 Apr 11
60	Install Pump (including interlock)	29 Apr 11
61	Install Centrifuge MCC (including E-stop relay panel connection)	29 Apr 11
63	Cheml Feed Room Commmissioning	2 May 11
64	AF Switchroom Commissioning	3 May 11
65	Commissioning of Tank Farm	5 May 11
66	Tank Farm - Plinth Excavation & Pour	5 May 11
67	AF ENERGISATION	5 May 11
70	Process Commissioning Bandscreen	5 May 11
71	Process Commissioning Intake Pumps	6 May 11
75	Construction Handover	6 May 11
76	Tank Farm Standard Process (Inst	8 May 11
79	Install Mixer Motor (including HU and interlock)	8 May 11
80	Install Mixer Motor	8 May 11
83	Pumps Electrical Activity in Tank Farm	13 May 11
87	WTG Commissioning standard process (pump, valves and instruments)	13 May 11
88	WTG Commissioning standard process (pump)	13 May 11
89	Commissioning standard process (pump)	13 May 11
90	WTG Commissioning standard process (valves and instruments)	13 May 11
91	Commissioning standard process (valves and instruments)	14 May 11
92	DN 1800 BV NORTH	27 May 11
93	Constuction-Commissioning Handover Process	27 May 11
94	Install supply to Vendor Panel	27 May 11
95	Install truck load-in panel (supply and E-stop connection)	27 May 11
96	Install RIO panel (two supplies)	31 May 11
97	Process Commissioning Bandscreen 1,2,3	16 Jun 11
98	CO2 E	16 Jun 11
99	Install supply to Chlorine Drum Heater	21 Jun 11
100	Bunding Services for Lime	22 Jun 11
101	Pit earthing	22 Jun 11
104	UF Cells MPB2	13 Jul 11
107	Install SDSS Pipework NE RO Gallery	18 Jul 11
108	Install SDSS Pipework NW RO Gallery	18 Jul 11
109	Install SDSS Pipework SW RO Gallery	18 Jul 11
111	Install SDSS Pipework SE RO Gallery	18 Jul 11
112	Tank Farm Process Commissioning 2	20 Jul 11
115	I&C - PIT (SDS) QA	1 Aug 11
116	I&C - PIT (PVC) QA	1 Aug 11
117	I&C - PDIT (SDS) QA	1 Aug 11
118	I&C - PDIT (PVC) QA	1 Aug 11
119	I&C - Analyser (SDS) QA	1 Aug 11
120	I&C - Analyser (PVC) QA	1 Aug 11
121	I&C - Flow Switch (SDS) QA	1 Aug 11
122	I&C - Flow Switch (PVC) QA	1 Aug 11
123	I&C - Pressure Switch (SDS) QA	1 Aug 11
124	I&C - Pressure Switch (PVC) QA	1 Aug 11
125	I&C - TIT (PVC) QA	1 Aug 11
126	I&C - Pressure Indicator QA	1 Aug 11
127	I&C - LIT QA	1 Aug 11
128	I&C - Level Switch QA	1 Aug 11
129	I&C - Electric Valve QA	1 Aug 11
130	I&C - Pn Valve QA	1 Aug 11
131	I&C - Pn Control Valve QA	1 Aug 11
132	I&C - Motorised Control Valve QA	2 Aug 11
133	I&C - Instrument Rack c/w Tundish QA	2 Aug 11
134	I&C - Pn Valve Box QA	2 Aug 11
135	I&C - Instrument Box QA	2 Aug 11
136	I&C - Profibus PA Box QA	2 Aug 11
137	I&C - Network Profi Box QA	2 Aug 11
138	I&C - Power Nice Box QA	2 Aug 11
139	I&C - RIO Panel QA	3 Aug 11
140	I&C - Manual Valve Limits QA	3 Aug 11
141	I&C - Pn Valve Install QA	3 Aug 11
142	I&C - RIO Panel Interconnect QA	4 Aug 11
143	I&C - Instrument Rack no Tundish QA	5 Aug 11
144	I&C - Profibus DP Active Terminator	6 Aug 11
145	I&C - RO Rack Completions	8 Aug 11
146	Standard RO rack punchlist	9 Aug 11
147	RO1 Racks - Mechanical Completion	11 Aug 11
149	CCC runaround	12 Aug 11
152	RO RACK PASS 1 SDSS INSTALLATION(2)	15 Aug 11
154	I&C - HPP Vibration Sensor QA	17 Aug 11
156	I&C - HP Pump Completions	18 Aug 11
157	RO Rack QA Documentation	19 Aug 11
163	UF Cell Completion	19 Aug 11
164	SP2 - Basic UF Cell Completion	20 Aug 11
168	E4,E5,F4,F5	20 Aug 11
169	Copy of RO Rack Pass 1 SDSS Installation	22 Aug 11
170	UFF D Train	22 Aug 11
172	UFF Cell D1 Mech Externals	23 Aug 11
173	SP2 UF Cell External Mech Construction Rev 2	24 Aug 11
175	RO Racks M&E Duration Basic	24 Aug 11
176	UF Cell Internals Installation	24 Aug 11
177	UFF Cell External - Electrical	24 Aug 11
178	SP2 Internal UFF Cell	24 Aug 11
180	ERD vent line modification	30 Aug 11
181	Install grating	30 Aug 11

Standard Process (Work) Library for Large Infrastructure Project

COMPUTER-AIDED PRODUCTION ENGINEERING

Through the application of computer-aided production engineering (CAPE), a library of standard work is developed, which is then used to plan, study, and improve work execution. It is a methodology that uses digital technologies to model, visualize, and simulate assumptions about how work processes are to be executed. What we're looking to do is validate the design using a 3D model, then understand how best to make the parts and assemble them, whether on-site or off-site.

Example

On another project in London, a very critical and complex scope of work needed to be executed at a rail station. Following is the story as documented by the project team, including how two major design issues were discovered during the CAPE process.

> In April 2004 the Fleet Sewer Team were tasked with removing the crash deck from the tunnel (installed on a previous possession), installing a Support Deck through the crown of the 'live' Thameslink railway tunnel, earthing, and bonding the 7 sections of the deck, installing 2 plate girders (weighing 65 & 105 Tonnes) and finally installing 5 M-Beams (weighing 30 tonnes each). All these works had to go through the scrutiny of the rail authorities and our Client Rail Link Engineering and had to be achieved during a 51 hour railway possession; the last realistically available before the Thameslink Blockade. Many thought this process and volume of work in the time available was impossible.
>
> The consequences of failing were unthinkable, The Train Operating Company delay costs being approximately

£400,000 for being just 4 hours late and potentially the far more costly incidence of the works having to be completed late during the main Blockade works.

The works were finished in just 37 hours! The Client was fully involved in the process and when it came to approval, had few questions or clarifications to ask of the team, thanks in part to the clarity/transparency provided by the 3D prototype. The team prepared for the works in detail, and by controlling their works in ProjectFlow SPS|PM, all the paperwork and approvals were in place by the time of the possession. Using ProjectFlow SPS|PM and the 3D prototype, the team understood the process and each team member knew their role. By producing the 3D prototype the team uncovered 2 design errors, that could have aborted the works had they not been solved early enough.

4D was then used to visualize how the works would be executed finding additional opportunities to compress the schedule duration. *If you can't build it digitally, you can't build it physically.*

After presenting this example at a Stanford CIFE conference content from the presentation was the basis of an article in *Engineering News Record* (*ENR*).

Finally, knowing the physical world can be far different from the digital realm, first-run studies (a term coined by Glenn Ballard) can be used to test our assumptions. In the general building sector, the requirement to construct various mock-ups (e.g., cladding mock-ups) creates perfect opportunities to undertake first-run studies.

For example, how long does it actually take to weld something? How

If you can't build it digitally, you can't build it physically.

much time is required to install a fastener? Some people balk at this notion. Really? You have to know how long it takes to install a fastener? Well, if you're installing fifty thousand of them, whether each one takes thirty seconds or two minutes makes a big difference when multiplied fifty thousand times. Understanding this kind of granular detail is essential.

CAPE also supports the environmental health and safety (EHS) risk mitigation framework by providing the means to identify any risk in a process and providing the framework to eliminate or guard against it. It is safe to say the majority of the effort now focuses on the behavior of craft people and the belief that changing their behavior is the key. This is in contrast to a more systematic approach to EHS, focused on (1) eliminating EHS risk, (2) guarding against risk—a guard on a saw, much like a physical control, (3) using personal protective equipment, and (4) educating, training, and assessing human behavior, which is all done during CAPE studies. In other words, EHS starts in the early part of a project in the design office (not on a construction site, where design decisions result in the need for less-than-safe operations).

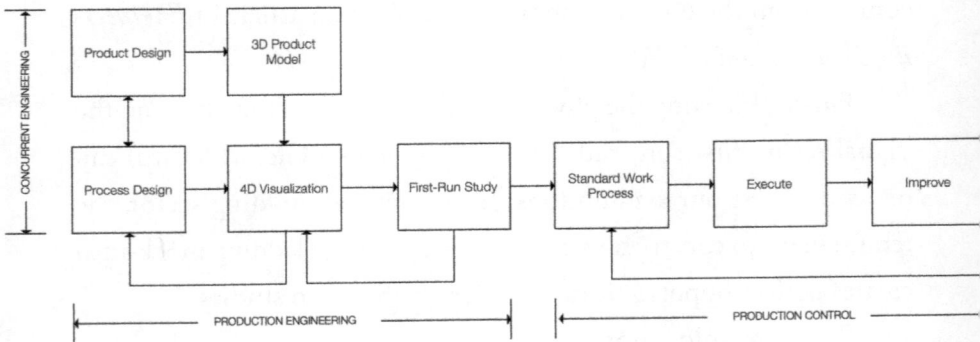

Computer-Aided Production Engineering Process

Example

This is exactly what happened while doing a large expansion joint installation at San Francisco International Airport. First, we created a 3D model, which indicated there existed problems with the architectural design. Specifically, the parapet was not high enough to accommodate the roof slope (really nothing to do with the expansion joint installation, but through the CAPE study we identified the problem). Once that was resolved, we then looked to understand the expansion joint installation process. One of our engineers created a process flow diagram, and we developed a 4D visualization. We made some modifications to the installation process, then set out to undertake the first-run study. Looking back at it now, we should have identified this in advance, but as we commenced the study, we began to discover the time to unpack the parts and discard the dunnage took as much time as the installation process (for future projects we learned to ask the vendor not to use excessive packaging).

It is very common to identify ways a product or asset can be improved during the process design/optimization phase of a CAPE study—or rather we feel confident that almost any construction operation will benefit from a CAPE study.

Example

For example, Pacific Contracting, which was to install louvers at the new Medical Center in Palo Alto, California, identified how a small change in the product design could reduce the cost and duration of installation by 90 percent while providing the owner with a far safer situation when maintenance needed to occur. If not done in a way that eliminated the risk of the louver being pushed in too far and falling out the other side, a serious safety incident could occur. This could happen during initial installation or even during some future maintenance operation.

This case also depicts how the organizational structure of the general building sector in the US results in unnecessary processing and rework associated with the technical submittal process. Hours of work and weeks of time taken so that the general contractor could transmit the construction documents to the subcontractors, who in return either create or have their suppliers create shop drawings, which are then submitted to the general contractor, who then submits to the architect (who may even send to the engineer). The general contractor or the construction manager is charged with coordinating the work prior to submitting to the architect. The architect and/or engineers review the submission and either place a stamp on the shop drawings stated they've been reviewed or ask questions on the shop drawings or outright reject them. If there exist questions or if the package is rejected, the process starts over.

Not only is standard work and standard process an output of CAPE but so, too, is work instruction. Not common in the construction industry, work instruction is of significant value when training new workers and when explaining how best to execute an operation to any worker, especially when working in a live or operating environment. As the reported labor shortage comes to fruition, ("reported," because if we use OS to get our shit together, we may not need as many workers as predicted) we, as an industry, may actually entice young people back into the profession by adopting emerging technology.

Though this example resulted in a positive outcome for all stakeholders, reduced risk of a serious safety incident, improved quality for the owner and faster execution of work for the general contractor, and more profit for the installation contractor, it clearly depicts the opportunity and need for production engineering. What it also shows is the extreme focus on product design (what work needs to be performed) with little if any understanding and effort placed on process design (how the work will be executed). As mentioned previously, this is no doubt a result of A&E firms avoiding means and methods as directed by their attorneys and insurance companies.

"The Contractor shall be solely responsible for, and have control over, construction means, methods, techniques, sequences and procedures and for coordinating all portions of the Work under the Contract, unless the Contract Documents give other specific instructions concerning these matters."[15]

Effective CAPE requires engagement and coordination of all parties responsible for the design of the product and the design of the work processes, including those involved in fabrication, transportation, installation, and others as needed to optimize potential product and process decisions. That said, significant opportunities exist to improve process design almost to the very end of a project.

Production engineering is based on identification of the bottleneck and use of science to resolve the issue. By *science* we mean operations science, materials science, and various fields of engineering. What we are doing is providing a methodology that leverages the competencies and capabilities of the project team and their know-how to optimize the life cycle of the asset through a rigorous process.

---

15   American Institute of Architects, Document A201, 2017, accessed July 6, 2023, https://content.aia.org/sites/default/files/2017-04/A201_2017%20sample%20 %28002%29.pdf.

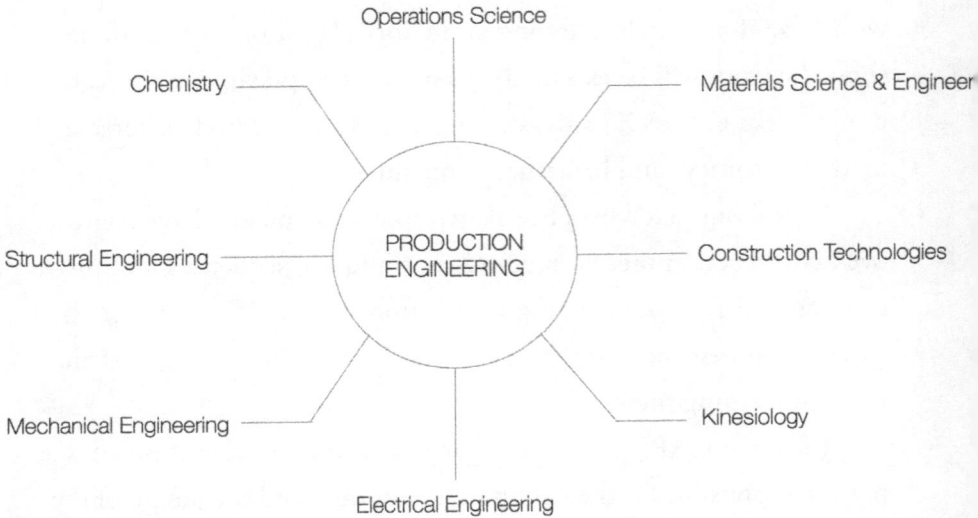

Operations Science

Chemistry

Materials Science & Engineer

Structural Engineering

PRODUCTION
ENGINEERING

Construction Technologies

Mechanical Engineering

Kinesiology

Electrical Engineering

Production Engineering Framework

In another example, while working with a precast yard to increase throughput and reduce cycle time, we identified various opportunities for improvement. These opportunities included (1) standardization of column sizing, (2) modification of the column molds to enable faster installation of the rebar cages and fit up of the mold, and (3) adding plasticizer to the concrete formulation to accelerate curing and minimize finishing.

By now you may be asking, How is CAPE different from design for manufacture and assembly (DfMA) or now design for life cycle (DfX)? The answer: DfX sets forth the standards for part counts, ease of making and assembly, etc., while CAPE provides the structured process for accomplishing these objectives.

Production engineering is ideally performed as early in the project delivery process as possible, when the ability to influence is the highest.

For instance, potential options available during early concept design are far more plentiful than when a mechanic is getting flummoxed at a construction site because he is unable to install the thing he needs to install. More on this in chapter 8, "Design Is Never Complete."

The following diagram sets forth a "maturity road map" for the use of standard work and PPC. It moves from documenting standard work to controlling standard work using LRM-based PPC followed by the integration of other technologies.

> **Potential options available during early concept design are far more plentiful than when a mechanic is getting flummoxed at a construction site.**

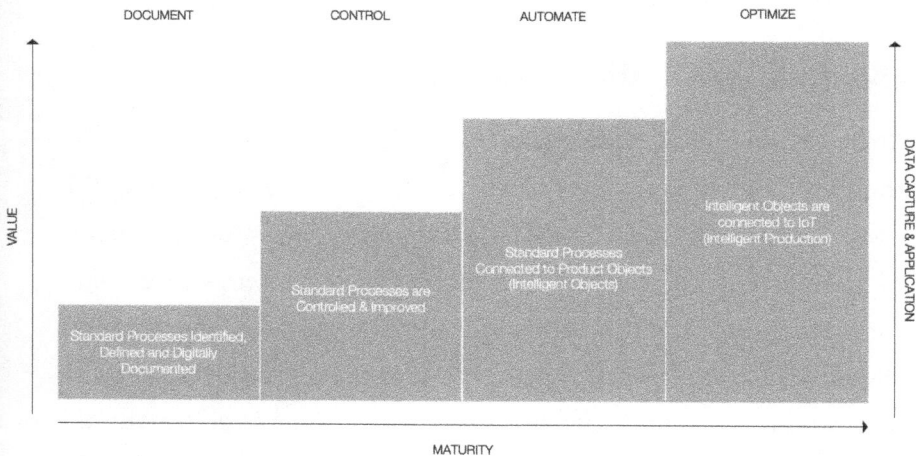

Copyright Strategic Project Solutions, Inc.

PPC AND PROJECT CONTROLS— WHAT IS THE RELATIONSHIP?

Many people inquire as to how PPC is related to project controls and the use of CPM. The answer is that it depends on your objectives. If you want to determine the robustness of a schedule, then port the tasks directly into the PPC system (see case study below). It is interesting to see how this strategy plays out. Project controls people almost always respond they are in the process of updating the schedule and will send it as soon as they are done with the updates. There are numerous projects that still have not provided the updated CPM schedule they promised. Don't get me wrong, these projects have been complete for years or even over a decade but still have no updated and accurate CPM schedule to work with!

If you want to compress the schedule, reduce cost, etc., then use the milestones in the project controls schedule as the basis of production control. And as stated above, since production rates drive schedule dates, PPC provides real-time project status (but this means planners and schedulers would need a new line of work—perhaps deployment and use of PPC?).

Example—Machinery Rebuild

There is an excellent case study that Boldt Company presented at the Fiatech conference in 2005 titled "The Result of Applying CPM 'Push' Scheduling to Lean 'Pull' Technology Tools" (the tool being SPS Production Manager). The work consisted of a series of rebuilds for a pulp and paper machine, and this was the first. As with any shutdown or turnaround, time is of the essence—an operating asset that produces products and the associated revenue is going offline. Interestingly, all sorts of executives tune into these projects. The VP of sales is eager to get back online, the CFO wants cash to be flowing in instead of out, and the chief legal counsel stands by ready to clean up the mess.

The project began using a CPM schedule created by a planner. The works commenced, and immediately the schedule "pushed out," potentially putting Boldt's client in an unhappy position. Augmented by production analytics related to reliability, Paul Reiser of Boldt, one of the foremost thinkers in lean and an early adopter of PPC, instantly recognized the situation.

As a demonstration, at one area of the project the use of the CPM was relaxed, and those responsible for executing the work were asked to prepare an updated production schedule. This production schedule was then used as the basis for production planning (which, by its design, is the engine to make the work flow through the process). The remainder of the project adopted the approach. The project was delivered on time, and the execs were very happy.

For a follow-on project, a third-party planner/scheduler once again created a CPM schedule. This time, to build upon the success of the previous experience, Boldt invited the project's site supervisors to provide input into the schedule. This effort resulted in the CPM being scrapped and an entire new schedule being produced by the superintendents and foreman. This schedule was then controlled using PPC. Interestingly, the project was executed using two shifts, with PPC also being the means for managing the handoff between shifts. Not surprising to most of us, the project set a world record, wherein prior best was twenty-five days, the goal was twenty-four days, and this project was done in twenty-three days with no safety incidents.

Per a senior vice president owner, "as the final single step in the project execution plan capital investment process, this rebuild represented an ideal outcome. The process of 'old press section out and complete new section in—all paper-to-paper in 23 days,' employed a NASCAR-style attention to detail, timing, and teamwork. This success is recognized by the entire paper industry as an outstanding piece of work."

SUPPLY FLOW CONTROL

Supply flow control (SFC) is a specific configuration of production control that is used to ensure all work necessary to engineer, fabricate, assemble off-site, transport, and deliver a defined set of materials, parts, subassemblies, or assemblies flows to a site (fabrication and/ or assembly shop or the actual installation site) in accordance with project objectives.

The best way to understand the role of SFC is to recognize there exists two flows occurring, whether in a fabrication shop or at a construction site, as depicted below. The first flow is the assembly process, which, as we know, is often represented by a Gannt or bar chart. It flows from left to right in time. The second flow is the supply of materials and parts that will arrive to the site, be processed, and assembled/ installed (if a fab shop will then be shipped to the customer).

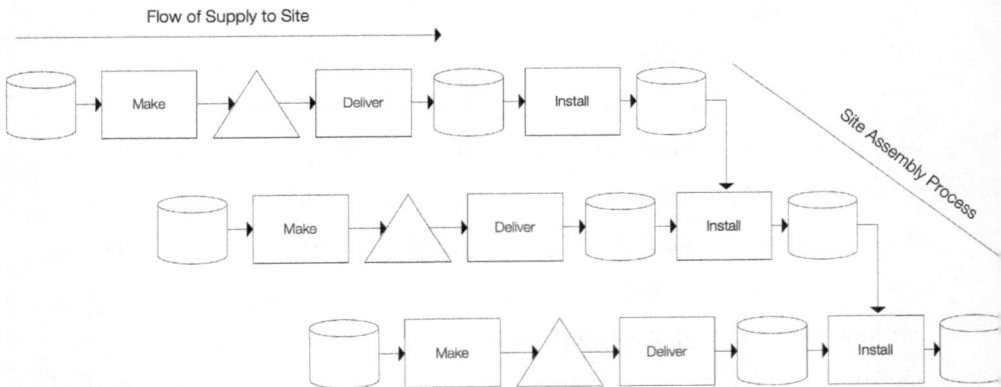

Copyright Strategic Project Solutions Inc.

Supply Flow and Assembly Flow

Most project professionals understand the site can only go as fast as it can be fed with information, materials, and parts. What is not well understood is how best to regulate the flow of stuff to a site—a flow based on getting the right materials, in the right quantity to the right place at the right time every time and at best value for the project.

For various reasons outlined earlier, project professionals prefer to be looking *at* it rather than *for* it, or to adopt a just-in-case versus a just-in-time approach. And with the current methods and support systems, who can blame them? Our observations over the past several decades conclude that the predominant model of supply flow control is a team of expeditors armed with phones and spreadsheets to track what is where and what needs to be done to get it to where it will be fabricated, assembled, or installed. Of course, there are also the more sophisticated operations that use RFID to locate shit they shouldn't have on-site and can't find.

Fortunately, PPM and operations science provide the framework for controlling (remember, *control* is not *controls*) the flow of supply, including synchronization of site demand with supply. Use of project production control on-site linked to an SFC solution incorporating the constant work in process or CONWIP control protocol enables control of WIP in the supply process and the associated use of capacity in the supply process and on-site.

> **PPM and operations science provide the framework for controlling the flow of supply.**

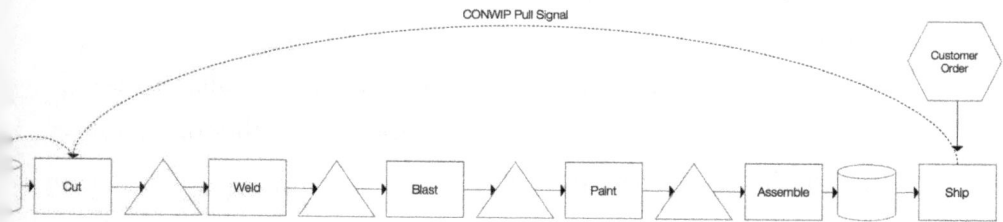

CONWIP Control Protocol

EFFECTIVE APPROACH TO SCHEDULING

To effectively schedule a project, one must recognize two fundamental elements of project delivery. First, as explained previously, a schedule is merely a prediction that, because of variability, *will* change. Second, that a schedule is the demand for a production system and that the production system is actually where the transformation of inputs (information, resources, energy, etc.) occurs. Wherein a schedule merely predicts what we want to have or what should happen, a production system dictates what *will* happen. Finally, unlike product development and software projects, construction projects are highly regulated. Numerous stakeholders determine what will happen in what sequence.

This brings into question concepts such as phase-gate processes and agile development approaches. Additionally, governing operations science principles are not understood or even recognized. This includes the fallacy of chasing productivity, the relationship between WIP, use of capacity (equipment, labor, and space), and project duration.

The advent of relational databases, increased computing power, mobile communications, and web-based distributed systems have enabled real time distributed planning capabilities. IoT-enabled control systems, AI, ML, and robotic process automation are furthering these capabilities and redefine how schedules are created (if they are even used).

So if current approaches do not work, how should project schedules be created? The following describes the methodology for production-based project scheduling.

```
                    ┌─────────────────┐
                    │ Milestone / Phase│
                    │    Schedule      │
                    └────────┬────────┘
                             │ Should Do
┌──────────────┐   ┌─────────────────┐   Can Do
│tandard Work  │──▶│ Production       │─────────┐
│Library       │   │ Schedule         │         │
└──────┬───────┘   └─────────────────┘         │
       │ How to Do                              │
┌──────────────┐   ┌─────────────────┐  Will Do ┌──────────────┐
│Computer Aided│   │ Production Plan  │────────▶│ Execute Work │
│oduction      │   └─────────────────┘         └──────────────┘
│Engineering   │
└──────────────┘
```

Did Do or Did Not Do

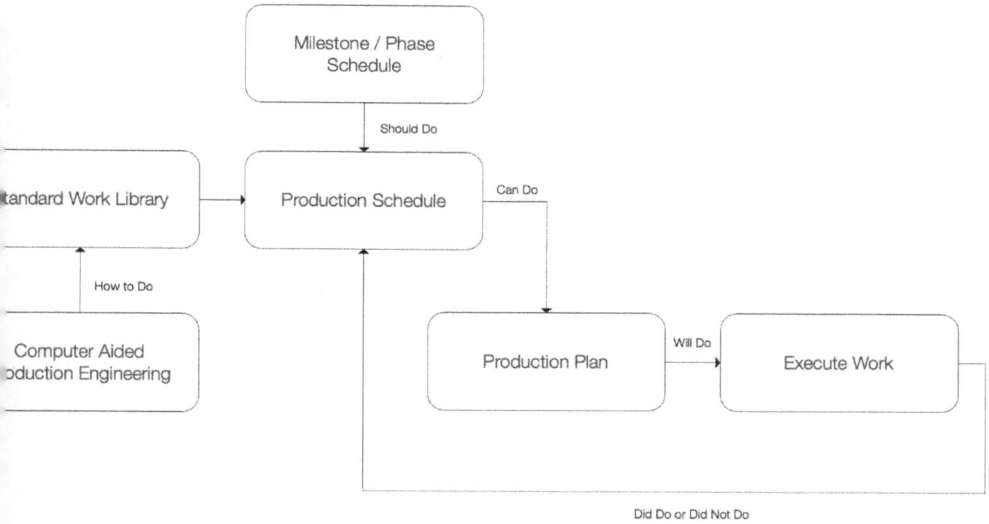

Strategic Project Solutions, Inc.

1. Identify and Define Master Milestones

Milestones are a specific scope of work that include conditions of satisfaction for handover to the downstream customer (versus conventional thinking where the focus is on dates—driven by the desire to measure and predict progress). Master milestones are those that represent the major phases of the project. Master schedules shall be limited to the least number of milestones needed to manage the project (see guidelines here). There are three primary types of master milestones: (1) regulatory, (2) technical, and (3) business. It is important to understand each type of milestone and its attributes.

Regulatory: Regulatory milestones are those related to regulatory requirements imposed on the project by local governments or other agencies. Examples include environmental impact studies, planning permissions, and permits including building, operation, and associated inspection.

Technical: Technical milestones are related to the design, delivery, shutdown, and start-up of the asset.

Business: Business milestones are self-imposed and relate to items such as when the asset needs to be operational or at full production; governance, including funding procedures/authorizations, whether internal or external; and policies related to master milestone transfer batch size (e.g., all engineering complete prior to physical production, etc.).

For the most part, regulatory milestones are altered through negotiation, technical milestones are changed through the application of science and engineering, and business milestones are revised through business planning processes.

Identification of master milestones is best done through interactive workshops where various stakeholders representing the regulatory, technical, and business aspects of the project are in attendance. During the workshop, each stakeholder presents the requirements set forth for each of the above three types of milestones. This documents and establishes the foundation for the phase-schedule basis of design and the project execution strategy.

2. **Develop Phase Schedule**

The phase schedule determines (1) the logic relationship between the milestones, (2) workflows required to achieve the conditions of satisfaction for each milestone, and (3) required completion dates for the work associated with the master milestones. In other words, the phase schedule sets forth the logical flow of master milestones as well as the work to achieve a master milestone.

It is important to recognize that there exists numerous contracting strategies and associated processes for delivery of a capital project, including design then bid and build, including lump-sum turn-key; design build; design-build bridging; EPCM; EPCI; etc. These strategies may also affect the required master milestones and the flow of work.

A – Engineer Complete then Make

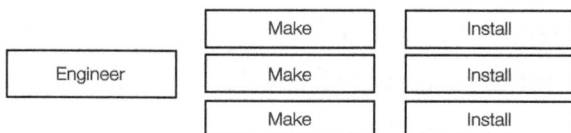

	Make	Install
Engineer	Make	Install
	Make	Install

B – Fast Track

Engineer		
	Make	
		Install
Engineer		
	Make	
		Install

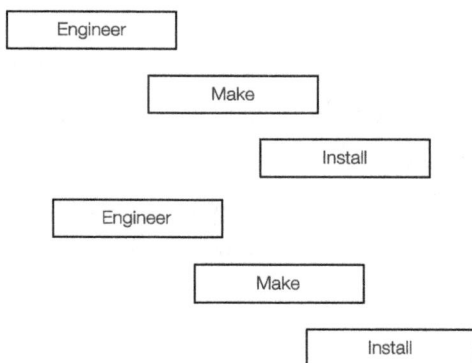

Figure—Examples of Project Delivery Strategies

There is also ongoing debate regarding whether engineering or construction should establish the flow of engineering work. Though many are of the opinion that construction should set the schedule or drive engineering work, the unintended consequences of this approach must be understood. Engineering has a technical flow it must follow, which is often far different from the flow of work in the fabrication shop or on the construction site.

Because of the extensive amount of engineered-to-order and made-to-order items associated with a major capital project, long-lead items must be identified as early in the project delivery process as possible.

Long-lead-time items are very antagonistic to efficient project production systems. Long leads force the need to forecast and result in the matching problem. Lead time shall never be accepted at face value and shall always be validated and compressed when possible.

Just as is done with the master milestone identification, phase schedules are developed using a cross-representation of the stakeholders involved in overseeing and executing the work. Phase schedules shall be developed using facility accepted by the operator as the primary milestone upon which to work back from. As with the definition of master milestones, workshops are the most effective means of developing phase schedules.

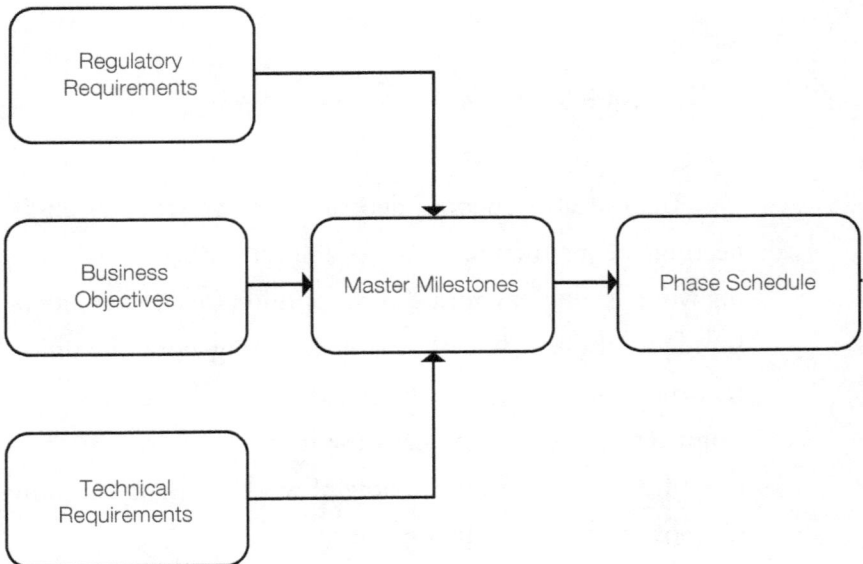

```
┌──────────────┐
│ Regulatory   │
│ Requirements │
└──────────────┘
                    ┌──────────────────┐      ┌──────────────┐
┌──────────────┐    │                  │      │              │
│ Business     │───▶│ Master Milestones│────▶│ Phase Schedule│
│ Objectives   │    │                  │      │              │
└──────────────┘    └──────────────────┘      └──────────────┘
┌──────────────┐
│ Technical    │
│ Requirements │
└──────────────┘
```

3. **Design Production System**

    As the phase schedule is being developed, the overall project production system and subordinate production systems should also be identified or designed. Early efforts include identification of potential sources of supply, including vendors and their location; lead time for supply; logistic routings; and so on.

    It is very important to understand that the production system delivers to the demand of the production schedule. Schedules provide the takt time in which the production system must respond, with the production system's response being in throughput and cycle time.

4. **Create Production Schedule**

    Ideally, all supply would be controlled with CONWIP or a pull signal. Production Schedules provide the demand forecast for the production system. But since long-lead items cannot be pulled, capacity contributors, including equipment and labor, must be understood and arranged for in advance, and a means of forecasting must be in place. Production schedules are not based on critical path method and are not used to predict and forecast progress. Production schedules use the last responsible moment scheduling technique. Technical sequence, transfer batch, and production-batch size are the key factors in creating a production schedule.

    Note: the transfer batch may be so large it decouples the need to follow a technical sequence.

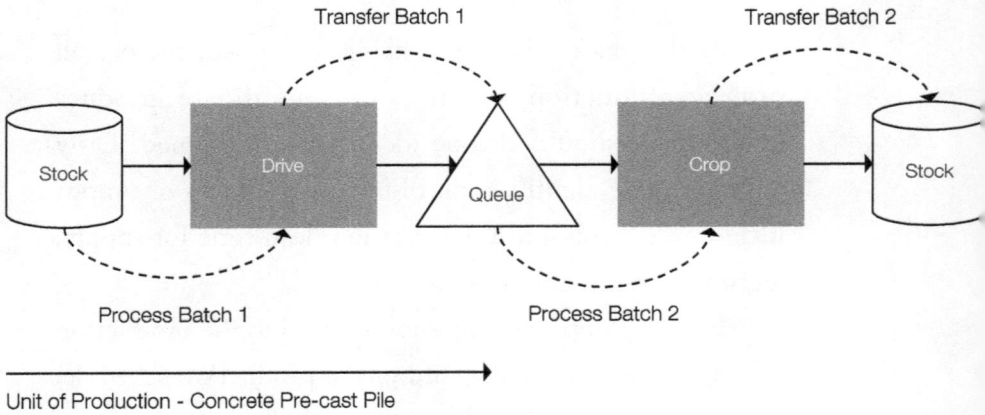

Transfer Batch 1

Transfer Batch 2

Process Batch 1

Process Batch 2

Unit of Production - Concrete Pre-cast Pile

5. **Control Production**

The most effective project production control solutions combine standard work and CONWIP, enabling synchronization of work through the production system. This results in effective control of WIP and use of capacity that, as you now know, reduces duration, cost, use of cash, and risk and improves quality.

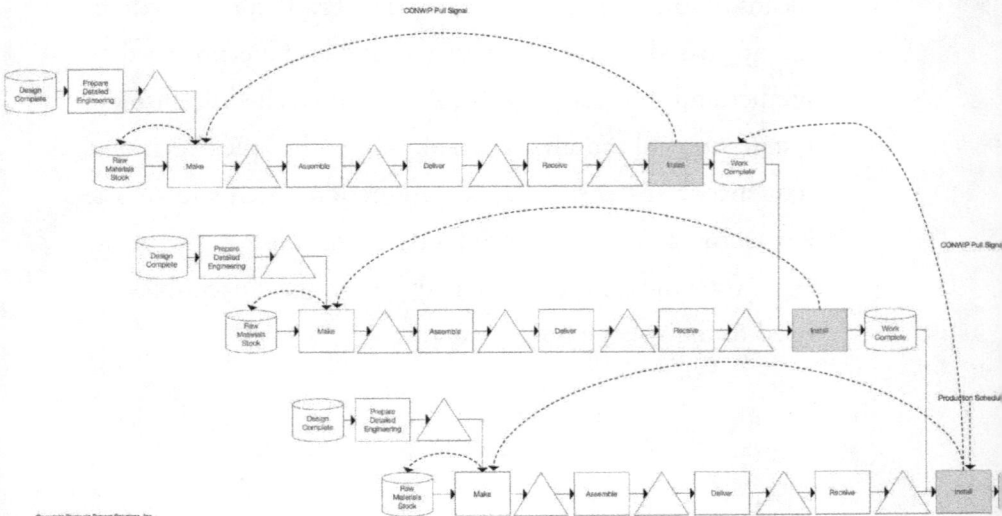

We often get asked about the potential value that can be derived from the implementation of project production management. The answer depends on (1) which element of PPM is being deployed and (2) the commercial arrangement between the owner and the contractors as depicted in the following diagram.

The more transactional the commercial agreement (e.g., fixed or lump sum), the more the value will go to the contractor. If the commercial model is relational, such as with a time-and-material-type agreement, the value will benefit the owner. That said, compression of schedule duration can benefit both the owner and contractor. The owner enjoys reduced outlay of cash and cost and the ability to realize revenue sooner, while at the same time the contractor often benefits from reduced labor, equipment, and overhead cost.

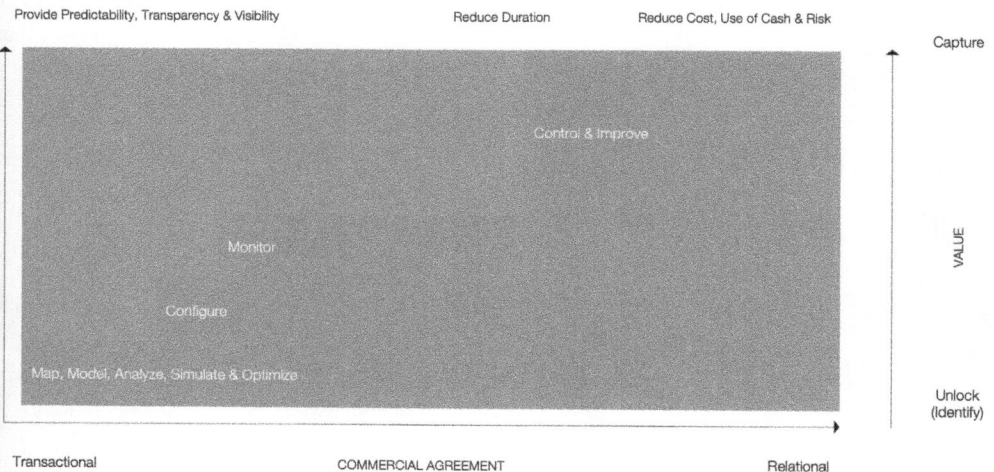

The next chapter will examine how all the shortfalls of Eras 1 and 2 come into play during the construction and commissioning phase of a capital project—and proposes a new framework for managing work at the construction site.

CHAPTER 5

HOW DOES PROJECT PRODUCTION MANAGEMENT IMPROVE SITE PERFORMANCE?

ALL THE SHORTFALLS of Eras 1 and 2 become evident during the construction and commissioning phase of a capital project.

Decisions about Era 2 work-breakdown structures and CPM schedules are made by planners who know little about how to execute the work, and if they do know, they are too far removed from it. Furthermore, Era 1 thinking about getting the most out of craft workers and the division between means and methods results in inaccurate or incomplete design information. Strategies based on using excessive amounts of materials or inventory buffers to avoid a situation in which workers lack necessary materials increases cost and risk while often shortening the time to fully complete the design.

Seeing the shortcomings of Eras 1 and 2 compound is like watching a sports team that doesn't have their shit together or an orchestra that didn't bother to rehearse—awful and sometimes embarrassing to witness.

Clearly, we're looking at the project management principle that states you can't deliver the whole project in one chunk; instead, you have to break it down into its component parts as needed. For example, if it's a building, you approach it not as the building as a whole but as the foundation, structure, interior, etc. But how you make decisions about the work breakdown, and the CPM schedules created by planners, affects actual physical production down the road.

Because of the overemphasis on administration (versus a focus on production), it is often said that planners of today are not even necessarily trained in construction, and they have a dim understanding of the interplay between various specialty contractors. The Era 2 planners view everything as large chunks of work in which they can measure and forecast progress, rather than integrating or coordinating the work.

Furthermore, the use of bar charts and CPM schedules hides the unnecessary WIP that accumulates during the project delivery process.

The division between means and methods results in inaccurate or incomplete design information. Designers, architects, and engineers have their hands tied by their insurance companies and their own internal risk policies, preventing them from doing anything related to establishing the means or process for executing work. Hence, the design information is often inaccurate because there is a disconnect between concept and execution, between what's being dreamed up in a remote air-conditioned office somewhere and what needs to be done on the ground, on-site.

The division between means and methods results in inaccurate or incomplete design information.

Meanwhile, questions abound around how to transition from

bulk construction to systems-based commissioning. Commissioning depends on systems: for example, we need to turn on the domestic water. Domestic water doesn't care if it's in the foundation, the basement, the first floor, or on the roof—it's the domestic water system. Now we have to think about how we take a whole system that is integrated with the facility.

Strategies are based on using excessive amounts of materials or inventory buffers to avoid a situation in which workers lack necessary materials increase cost and risk.

This is compounded by project professionals' stubborn belief—a legacy of Taylorism—that the issues lie in motivating workers to work harder. For instance, the fallacious thinking of a project engineer or project manager leads them to believe that craft workers will happily undertake rework as needed, because extra hours equal extra money for those workers. But in my several decades in this business, I have never met a single craft worker who enjoyed anything about rework, regardless of the root cause of the rework.

It is indisputable that the current approach is not working. My company's data on millions of tasks that we track in our software shows that only 53 percent of what goes on the next day's plan is completed. That means that despite the exhaustive attention poured into scheduling and administration, if we put ten things on the plan at 4:00 p.m. today, only five will get completed tomorrow. That's a startling statistic. It's even worse for longer timelines. If we forecast out to a week, that rate falls to less than 20 percent. Two weeks out? We can expect to complete only one task out of ten.

Using current approaches, we can't effectively predict what we'll do tomorrow, next week, or the week after. How do we expect to predict what will happen a year or more out?

DIFFERING PERSPECTIVES

Another factor is the different perspective or even challenge faced by the owner or construction management firm responsible for delivery of a project and a specialty contractor that only is responsible for a specific portion of the project (electrical engineering, steel erection, etc.).

Whereas owners and construction managers see the focus being the work of a specific project, service providers such as A&E firms, fabricators, and contractors see a project being just one event in a larger portfolio of work. Service providers allocate capacity across their entire portfolio. This is why CPM is of limited use to design firms, fabricators, and specialty contractors. As stated earlier, from the perspective of the service provider or product supplier, a bar chart schedule is the demand side of the equation. The supply side is what capacity the service provider or product supplier has and how they plan to allocate it for a given project in the context of their overall business plan.

Service providers understand that dedicating capacity to a specific project almost always results in loss of total capacity. Sure, the service provider can get paid for the resources, but the resources may be needed for technical purposes or may be able to deliver increased revenue for another client. Workflow and resource based schedules depict how the same schedules being seen from the owner or construction manager's viewpoint can be much different from that of a service provider. The logical sequence of design or construction is not the logical use of resources.

These schedules also depict a workflow (technical sequence of work)—that is, the perspective of the owner and construction manager—versus the resource-based view for the same work—that is, how a service provider sees the same work. There are often gaps in workflow in the resource-based schedules. Does the service provider maintain the capacity for one project? If so, does the owner compensate the service provider? If not, does the service provider absorb the cost or does the service provider allocate the capacity elsewhere?

This is precisely why Fondahl stated that resources will dictate the project schedule. If you want to know if a service provider will meet a date, then you must understand how they plan to allocate their capacity, whether in a fabrication shop and/or at a construction site, one must look at how they allocate across their portfolio of work—exactly what a PSO effort will tell us and what PPC addresses.

While visiting a very large and sophisticated shipyard in Korea for one of our energy customers, we asked an executive at the yard how they manage the work of the yard. The executive stated, "We do three things: (1) we have a long-term strategic schedule that we use for demand and capacity planning (but we can't show that to you), (2) we have the schedule you make us update to show progress and get paid (otherwise it is not much use to us), and (3) we have the production control schedule we use to control the resources in the yard—which we cannot show you, either, as it is confidential to our operations." Our customer had almost one hundred people in the yard monitoring the progress and forecast updates of their schedule and didn't even realize the production control schedule set forth what would happen, even for their project, which was one of several in the yard at the time.

What is needed is a framework upon which to understand production along with methods and tools to model, optimize, control, and improve construction-site production.

The 4-5-3 framework described in chapter 4 does just that, it provides the means for understanding and influencing construction-site production. Through 4-5-3 we can more effectively understand and positively influence work in the fabrication shop and at the construction site. But first, let's understand the complexity of planning work on a construction site (and I don't mean scheduling but the actual planning done by superintendents, general foremen, foremen, and even craft workers). But before that, let's understand the challenges associated with planning work in a fabrication facility or on a construction site.

THE PRODUCTION PLANNING CHALLENGE

Managing physical production, whether in a fabrication shop or at a construction site, is full of challenges. Maintaining a flow of work requires a flow of inbound information and materials, effective allocation of capacity contributors (equipment, labor, and tools) and management of variability. And to be clear, "moving the work off-site" (what Mark Reynolds, CEO of Mace, terms *builders in sheds*) does not mitigate these challenges. Unlike processing and line flow production, where the item being made flows "by" the process centers (machines, robots, and people), production at a construction site and even a fabrication shop is based on the workers working in and around the items being made or constructed.

When a supervisor or even a craft worker puts together a work plan or production plan, they're engaging with a design problem, one that is resolved through deductive reasoning. They have to figure out (1) what needs to be done and (2) whether they have what they need to do it. Unlike planners creating bar charts and CPM schedules, they cannot kick the can down the road and say that it is site supervision's problem. They are the site supervision and have no choice but to deal with the matching problem described in the following paragraphs.

First, specifically, they must understand what policies, procedures, and objectives they are working toward, whether that involves complying with safety requirements or devising a policy around the ratio of journeymen to apprentices, or some company-imposed benchmark that says, "You must install five of X within a single day."

The next part of the equation addresses whether they have the materials and information they need to do the work. And they must determine whether the essential preceding

> When a supervisor or even a craft worker puts together a work plan or production plan, they're engaging with a design problem, one that is resolved through deductive reasoning.

work has been completed. This applies in myriad ways: if the project is composed of interlocking production systems, A must precede B before B can take place. If I have to install a pipe spool and I need to go up on a scaffold, has the scaffolding been installed? And do I have the equipment, labor, and space to do the work?

That is something supervisors (whether foremen or general superintendents) are concerned with day in, day out: Do I have what I need to do the work? And do I have it in the right sequence?

The concept is elementary enough, but actually resolving this problem on a capital project can get staggeringly complex.

To effectively plan the work at the production level (meaning the actual doing of the work), we must decipher between resources that are dynamic, such as labor, equipment and space, and resources such as information, materials, and parts that should be in stock (in the appropriate amount of stock). Preceding operations (often executed by others) along with availability of equipment, labor and space are very dynamic, with availability often not known much more than hours in advance of being needed.

Therefore, to address the "matching" problem, capacity contributors can only be allocated the evening before or the morning of the operation or task (the capacity contributors can be held as buffers, but, as always, there is a cost to this strategy). This means monthly or even weekly project controls schedules are of limited use to those responsible for managing work at the point of assembly.

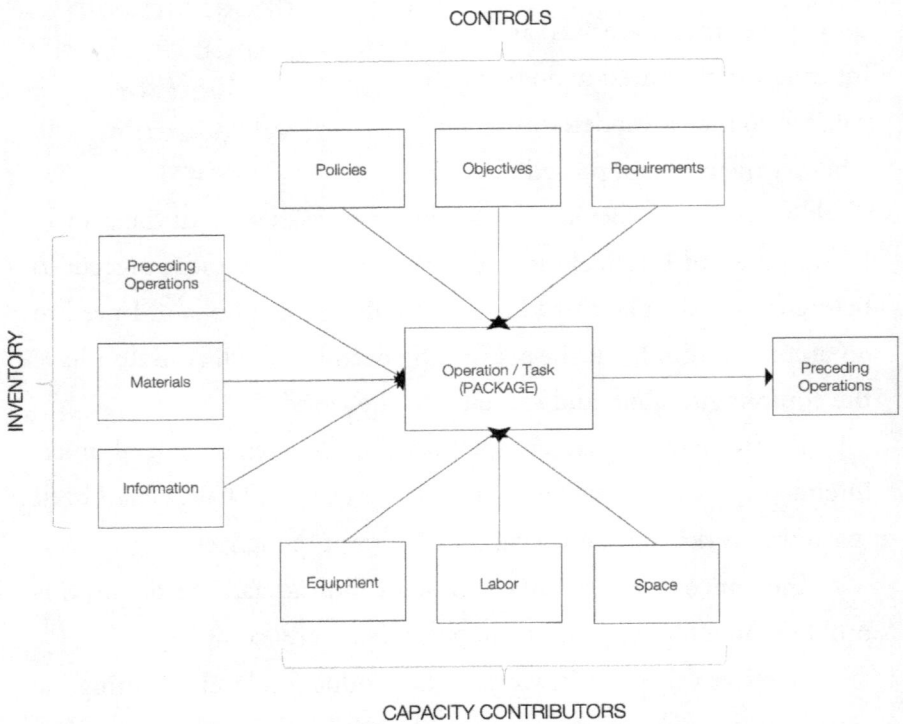

Copyright Project Production Institute

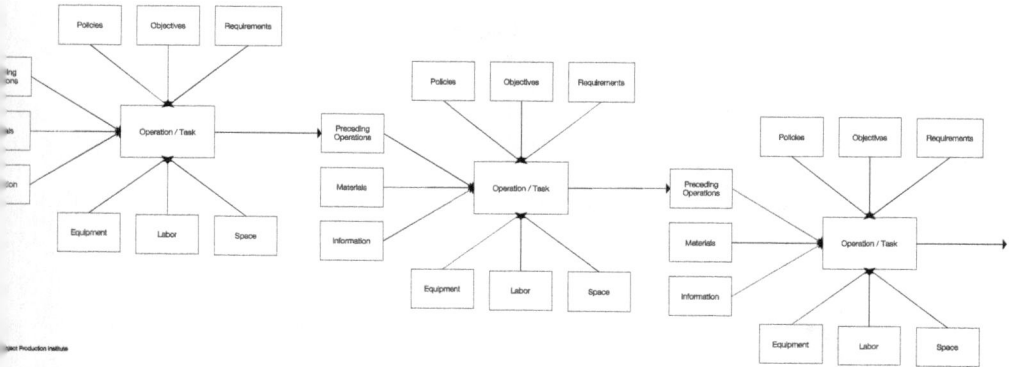

INDUSTRIALIZED CONSTRUCTION— WILL IT SOLVE THE PROBLEM?

To address these challenges, owners and contractors are working to industrialize construction, including standardizing design, moving assembly into what promises to be safer and more productive shop environments, and/or adopting various means of automation. These efforts consist of anything from fabricating and assembling subassemblies, such as a toilet and its piping, to more complete assemblies, such as large modules for an energy project or bathroom modules for dorms or hospitals.

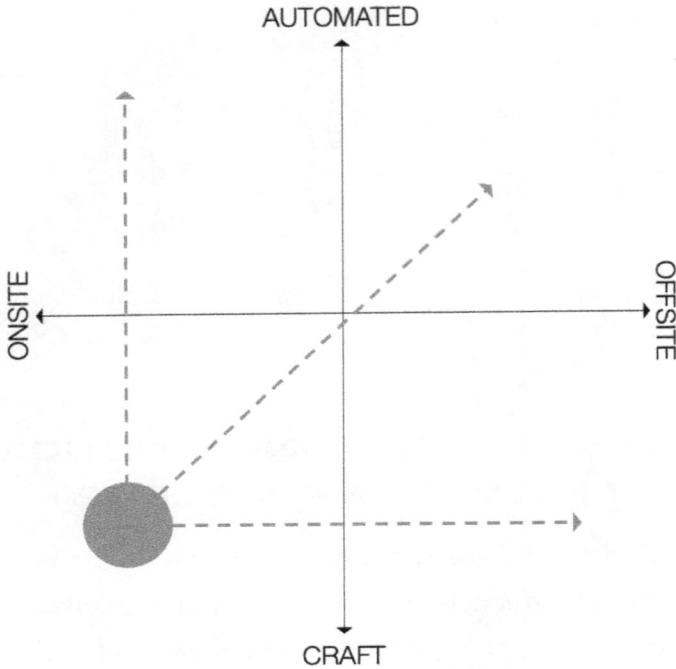

Copyright Project Production Institute

Industrialized construction can best be understood as the inter-section between construction and manufacturing. The big idea is to take the best from both sectors. But the fact remains: as long as the focus is on administration versus production, investment in these approaches and technologies will not deliver their full potential and, in worst case, might result in unintended consequences.

As mentioned in the introduction of this book, moving work off-site is no simple task and is fraught with complexity and risk, while at the same time there exists a reason as to why standard designs are not for everyone. Katerra and Westinghouse prove this point.

Sure, there is the benefit of doing site work concurrently with fabrication and assembly of modules, but the increased complexity associated with managing tolerances in space and time, and logistic challenges–including the need to ship air and provide temporary support during transport—often result in the need for more advanced means of rigging and moving stuff around the construction site and offset the benefit. It has also spawned the concept of system completion. Since an entire plant or high-rise building cannot be shipped at once, there remains the need to connect each module structurally, mechanically, electrically, etc. The benefit of the large batch in the form of a module is offset by what appears to be endless systems-completion work, whether by design or as the result of "ship loose" items (work supposed to be completed in the shop but not).

Like a good friend told us a while back, "We are getting the time out, but the cost is killing us." What this tells us is that the focus remains on standardization and off-site assembly without any understanding of operations science, including the implications of various buffering strategies related to capacity and inventory. Inventory and capacity buffers are being used to compress time (remember, inventory is the proxy for time, and capacity is used to build inventory, whether WIP or finished goods—which are inbound stocks to the downstream customer). But it all comes at a cost.

This is validated by a recent talk James Choo and I gave at a conference focused on moving work off-site. It was staggering to see an audience of professionals eager to adopt a manufacturing approach to construction and their businesses. We laid out the difference from being a fabricator using an engineered-to-order service model to a configured-to-order, made-to-order, or made-to-stock product supplier: how the business model is different, how getting paid for engineering morphs into investing in product development, and so on.

Service Provider

Product Supplier

$$\longleftrightarrow$$

Cost + Fee = Price
Engineer / Configure to Order
Demand Creation
Capacity Management

Price – Cost = Profit
Make to Order / Stock
Demand Creation / Management
Inventory Management

These soon-to-be ex-contractors and now manufacturers had no idea what we were talking about or why they should care, other than our buddy Bob. Bob said, "I tried this shifting to manufacturing shit and it's fucking expensive: fucking patents, testing, approvals and production equipment, it goes on and on." Most interesting was the lack of interest anyone in the audience had about how operations science is fundamental to operating a manufacturing-type operation.

Investment

Low Initial   High Initial

Design / Integrator

Service

                    Fabrication

                                            Capacity Utilization

                              Manufacturing

Product

                                            Inventory

                                    Processing

Cost + Profit = Price   Price – Cost = Profit

Financial Model

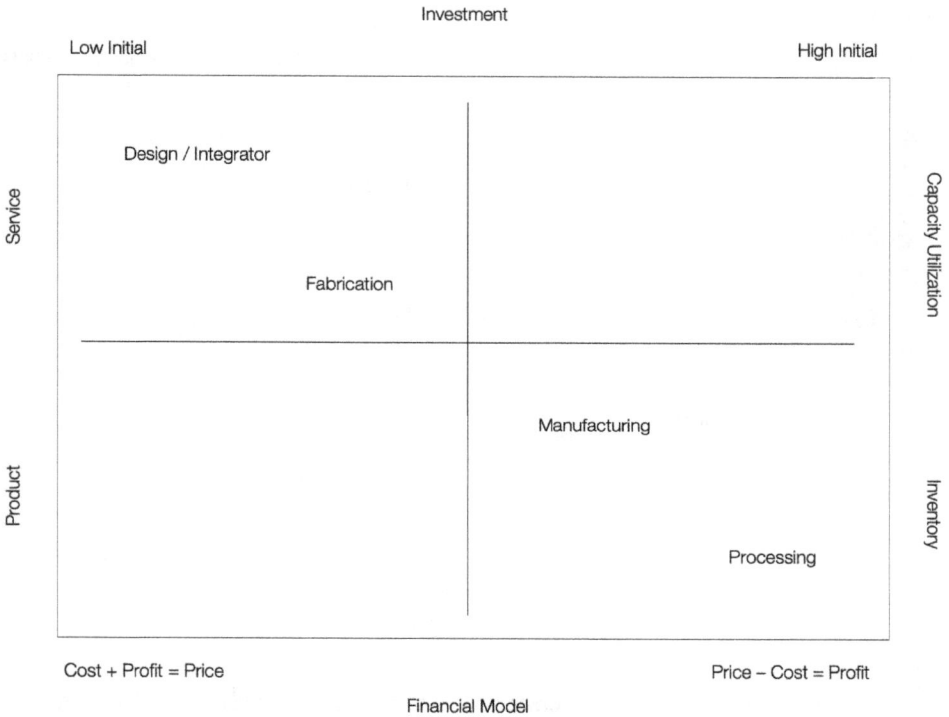

Copyright Project Production Institute

The result is lots of discussion and trial and error among industry professionals looking to understand which approach is best: create a kit of parts, a single-trade assembly (meaning one company makes a subassembly for the work of their trade), multitrade assembly (multiple trades or even companies, such as electrical and plumbing, are incorporated into a subassembly), and finally, multitrade volumetric, which include examples such as an entire bathroom module for a hospital or a topside for an offshore platform.

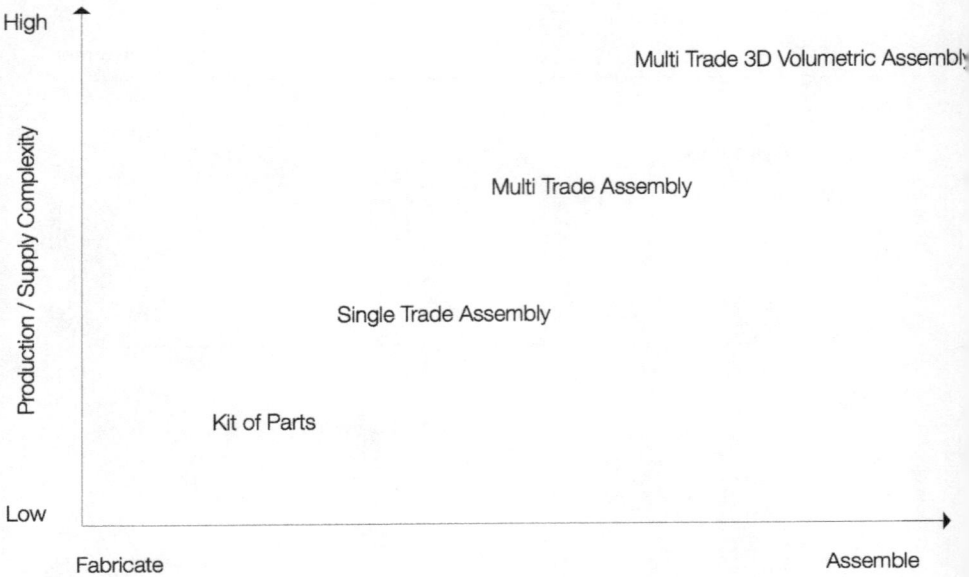

Copyright Project Production Institute

It is important to understand that the questions being asked are really how and where to best do the work. Again, these questions are best answered through the application of OS through a PSO study and use of CAPE to support product and production system design, as well as PPC and SFC to ensure the production system behaves as planned. Doing so results in saving incredible amounts of resources and perhaps even avoiding business failure.

How to do What, Where?

| Mine / Recycle | Process | Transport | Fabricate / Manufacture | Deliver | Install |

Onsite, Offsite, Craft, Automate?

Copyright Project Production Institute

We propose that assembling or even fabricating off-site is not always the best decision. A precast concrete company (an early customer of SPS) was purchasing rebar cut and bent by a supplier, which the supplier would deliver to the precast yard. The precast company would then assemble the rebar into cages for the precast elements. This seemed to be totally logical—work was moved off-site. But the unintended consequences were significant from a cost, time, use of cash, quality, and, most importantly, safety perspective. Interestingly, all deliveries were coming in two-ton bundles. Different pieces for different projects were bundled together; we thought this strange and decided to visit the supplier. While in the shop, we looked up and saw the answer: the maximum amount the overhead crane could lift was two tons. Therefore, all work was planned, fabricated, and bundled in two-ton batches (regardless of what was needed or best based on the situation).

The work at the precast company was the first-run study for Heathrow Terminal 5 (T5), which, upon understanding the value a raw materials stock and capacity buffer offer, adopted a similar approach. With 80-percent-plus rebar to be fabricated and assembled off-site but near the site in dedicated logistics centers, T5 was the largest rebar operation in Europe at the time. The lessons learned from the prefabrication effort were then applied to the T5 rebar production system.

This brings us to how Ford's use of knockdown kits and Ikea's use of flat packs may provide some insight into how best to approach moving work off-site, along with other manufacturing strategies related to bill of materials accuracy, sequencing production, and presentation of parts based on assembly requirements, and in process quality processes.

Construction professionals often look to use packages of work or bag-and-tag materials and equipment that can be delivered to the project site. Although this is an excellent idea, if the packages are created too far in advance, the nasty matching problem arrives on the scene. Because of variability (remember, only five out of ten items on a plan get completed daily, and at a week out is often two items or less), predetermined packages of work are not aligned with ever-changing site conditions.

At a chemical plant in Europe, the contractor went to great lengths to "package" the materials and deliver them to the site. Unfortunately, the contents of each package and the sequence in which the packages were delivered were not aligned with the actual situation on-site. As is common, the work on-site was not following the schedule used as the basis to kit the materials in advance. This made for extensive rework. Again, well-intended actions with unintended consequences.

COMMISSIONING

Commissioning is the point in the project life cycle wherein the owner begins to take possession of the asset. The owner's operations team along with the original equipment manufacturers (OEMs) and construction companies collaborate in a process of inspection, testing, and start-up of the various equipment and sometimes systems.

These processes follow a very specific technical sequence that is well documented by the OEM and the engineer of record. That said, our observation is that these processes are most often tracked using check sheets rather than controlling them as a production process wherein the detailed commissioning procedures form the basis of standard work.

One of the first people we talk with when visiting a construction site is the commissioning manager, even if the project is in the design phase. As the last person in the chain, the commissioning manager has an excellent view as to how the project will be delivered. They see the path of construction, the sequence of work, and often the massive transfer batch between bulk construction that is area based and commissioning that is systems based.

The desire to show progress (a.k.a. burn and earn) drives project professionals to adopt a "complete any work that can be completed" strategy. Decisions are made to complete work with or without the necessary materials and parts or to even use temporary parts to maintain schedule.

In some projects, people will do anything necessary to maintain progress with chain hoists as the solution for supporting the pipe spools. Known in the industry as "show pipe" this creates serious challenges when it comes time to commission a facility. The use of work-breakdown structures that are designed to support area-based bulk construction make the problem worse.

In the next chapter, we get into the thorny issue of supply and look critically at the prevailing notion that it's always better to have materials waiting on workers rather than workers waiting on materials. But there's a better alternative to both approaches—a just-in-time model fit for Era 3.

CHAPTER 6

SUPPLY: FROM JUST IN CASE TO JUST IN TIME

WHEN IT COMES TO MANAGING materials for a construction project, most construction professionals think it's better to "look at it then look for it." And why not? Suppliers get paid for materials, specialty contractors earn a markup on the cost, Era 2 planners can mark off the expenditures as "progress" based on earned value analysis, and the owner is happy to see progress being made.

The obsession with increasing worker productivity drives construction professionals to do whatever it takes to get materials on-site as soon as possible. The trend to offshore manufacturing and the associated supply chain disruptions, whether industrial action at ports or, more recently, COVID-related supply disruptions, further incentivize construction professionals to get materials to site or close to site as soon as possible, lest there be a shortage somewhere and "workers can't work." And they either don't know this is problematic or don't care, because (1) it's not their money—it's the owner who is ultimately buying materials, (2) the specialty contractors and product suppliers are still getting paid, and (3) the people tracking project progress can count the cost as a benchmark of progress. Moreover, there is a fundamental misunderstanding around productivity versus production throughput—high productivity may result in less work getting done and schedule duration being extended.

Project leaders are so well versed in this approach that they are willing to invest the cost and cash needed to handle, hold, and preserve materials (including permanent equipment), assuming the risk of damage, theft, and obsolescence due to design changes as simply the cost of doing business. Ninety percent of construction professionals operate under the mantra of "materials waiting on people is always better than people waiting on materials."

> There is a fundamental misunderstanding around productivity versus production throughput—high productivity may result in less work getting done and schedule duration being extended.

At a new chemical plant, 80 percent of structural steel for the whole project had been delivered to site, and over the next sixteen weeks 150,000 pipe spools and 7,500 underground spools would be delivered to site. Laydown on-site was near capacity for some items, with overall laydown capacity utilization at 83 percent, with 26 percent of anticipated materials received on-site. None of these items were installed as the project was in the site-civil-works phase at the time. At the same time, the project was experiencing a 40 percent to 50 percent planned-start-to-planned-finish success ratio of schedule dates based on a thirty-day lookahead window (the low reliability in achieving schedule dates would only increase the WIP of inbound materials as work got delayed and deliveries continued). The CM firm was working to convince the owner they needed more laydown space and five hundred additional trailers to accommodate the inventory—and it was only a matter of time before the CM would request additional material handling equipment and labor to manage all this. A show of force, no doubt! But at what cost?

The desire to commence production and delivery of materials too early is a potent cocktail of unnecessary use of resources, and the result is a crippling hangover of wasted time, squandered cash, and needlessly elevated risk. Present initiatives to industrialize construction by moving work off-site may actually exacerbate the problem because it just causes even more WIP to build up, now at two sites instead of one. Well-intended actions with unintended consequences.

Cost of Inventory

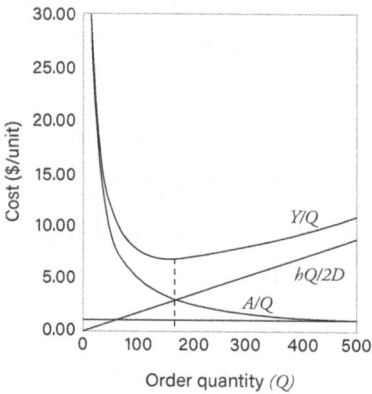

D = demand rate (in unites per year)
c = unit production cost, not counting setup or inventory costs (in dollars per unit)
A = fixed setup (ordering) cost to produce (purchase) a lot (in dollars)
h = holding cost (in dollars per unit per year); if the holding cost consists entirely of interest on money tied up in inventory, then h = ic, where i is the annual interest rate
Q = lost size (in units); this is the decision variable

$$Y(Q) = \frac{hQ}{2} + \frac{AD}{Q} + cD$$

The answer is to move from just in case to just in time. But what does *just in time* mean?

CASE STUDY

We once visited a large project where residential homes were being purchased and razed to make room for storage of structural steel, pipe spools, and other permanent plant equipment. When we arrived on-site, we observed another project with endless rows of flatbed trailers holding structural steel, pipe spools, and other materials needed to complete the project. To make additional laydown space, the project was purchasing homes in the area and razing them. The construction management firm was very proud of the site-materials management system they designed and deployed. The system included (1) getting all the materials out to site as early as possible; (2) all materials staged on 45-foot trailers, with smaller items bagged and tagged, to enable ease of movement; and (3) various tools in place, including a laydown space allocation tool on a white board, spreadsheet-based trackers, and access to an automated bill of materials from the 3D model.

Installation work packages (IWPs) were defined a minimum of 12 weeks in advance, followed by the bill of materials (BoM) for the IWPs six weeks in advance of need on-site. IWPs were kitted at three weeks in advance of need on-site and delivered to site one week in advance of the installation date.

Lots of haranguing around between the site and the logistics team was occurring, including the logistics team asking the site install teams what they needed weeks in advance, with the site responding, "Just send anything you have. We will take anything; we just need to burn hours." The variation in the planning of what the site would need and what the site actually needed resulted in the matching problem coming into play.

As stated earlier in the book, the ability to predict what exactly will be needed on what date is extremely low and gets lower the further out you plan. To address this issue, IWPs consisted of between

1,500 and 3,000 hours per package for a single trade. With this much work associated with each IWP, and often needing more than one IWP to maintain the path of work, the WIP on-site exploded. And of course so did cost, time, and use of cash. Needless to say, the system was not designed to address the inevitable variability that would affect the production system, creating a matching problem between what was prepared for shipment and what was actually needed.

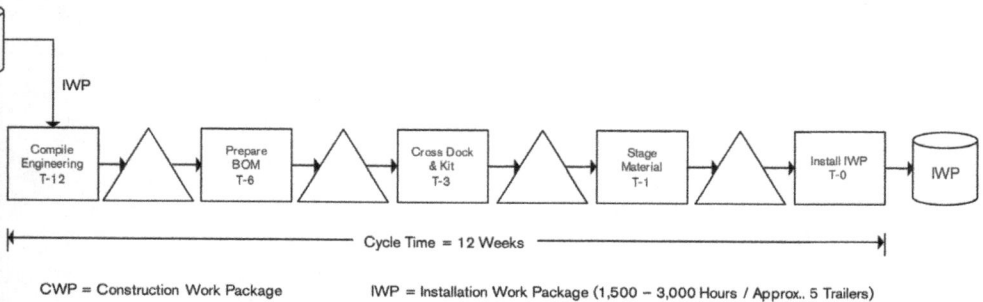

CWP = Construction Work Package   IWP = Installation Work Package (1,500 – 3,000 Hours / Approx.. 5 Trailers)

Strategic Project Solutions, Inc.

What we did to solve the problem was totally counterintuitive to what the project team deemed to be the answer.

1. We introduced the concept of a production package, which was to be one shift's worth of work for a crew.

2. We lowered the time from prepare BoM to install from six weeks to one week. Of course, anyone with a fundamental understanding of OS knows that the goal was to mitigate the matching problem.

3. We also removed the scheduling effort for the supply element and replaced it with a CONWIP signal from site.

4. Specific policies and measurements were put in place for every step in the process. See the following diagram.

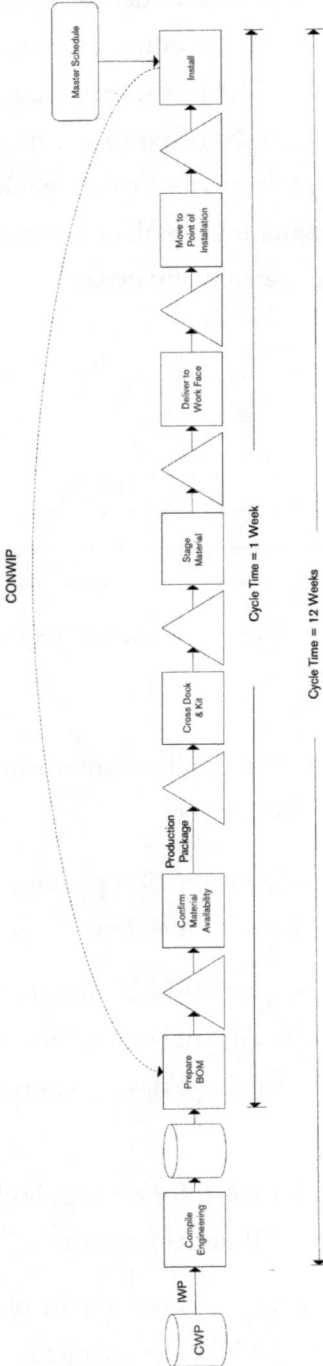

CONWIP

Master Schedule

CWP → Compile Engineering → Prepare BOM → Confirm Material Availability → **Production Package** → Cross Dock & Kit → Stage Material → Deliver to Work Face → Move to Point of Installation → Install

Cycle Time = 1 Week

Cycle Time = 12 Weeks

CWP = Construction Work Package   IWP = Installation Work Package (1,500 – 3,000 Hours / Approx. 5 Trailers)   PP = Production Package (1 Shift for 1 Crew)

1. Initiate production of a PP from the worksite (CONWIP)
2. Contain no more than one day of field installation work in a PP
3. Include all necessary technical information in a PP
4. Check for accuracy / quality of PP's at the logistic center and notify of issue
5. Kit today what will be delivered tomorrow
6. Optimize offloading sequence and parts presentation to the extent possible
9. Optimize kitting / delivery capacity within the day – no more
10. Deliver today what will be installed tomorrow

11. Ensure PP is sound when receiving at workface
12. Use trailers for staging when possible
13. Offload from trailer to Point of Installation (POI) when possible
14. Remove from site what is not consumed (waste, dunnage, rejected PP's, etc.) daily
15. Inspect and accept completed work daily

Copyright Strategic Project Solutions, Inc.

In so doing, we enabled the supply to match the site demand. Site got what they needed while supply only needed to do the minimum to keep the site supplied. These two efforts significantly reduced the WIP.

I often wonder what those homeowners thought once they received the cash, only to see their homes leveled by excavators.

DIAGNOSING THE LEAD TIME PROBLEM

Project teams often separate project materials or supplies into three categories:

1. **long-lead specialty items**, such as a compressor or chiller unit, or, if you're building a house, custom cabinetry or custom stone,

2. **bulk materials**, like aggregates or rebar; stuff that's generally readily available (recent COVID-19 disruptions notwithstanding), or

3. **consumables**, for example, welding rod or tape—you use it and it's gone. It's not really part of the finished product per se; it's consumed during the process.

Because of the risk of delay in receiving long-lead items and the implications to the schedule, these are of particular concern to project teams. They're ordered far in advance of the completion of design (which contributes to the aforementioned problem of obsolescence). In fact, design is often expedited to get a jump on ordering long-lead items. It is often said design is late, but my colleague James Choo always asks, "Are you sure you are not starting procurement too early?"

The current strategy of "get everything out as fast as possible" has a temporal, financial, and productive cost. It is anything but free. We don't really measure the real cost of the strategy, and worse, we incentivize it because too many people involved are rewarded for it, whether in the form of "progress" for the Era 2 managers who are tracking expenses or cash flow for the guys who provide the material.

> **The current strategy of "get everything out as fast as possible" has a temporal, financial, and productive cost.**

The owner suffers for it. But beyond that, we all suffer, because this strategy is a big reason why the industry is so broken, unwieldy, unreliable, and slow.

Lead time is the time from placing the order to when the customer gets it. The cycle time is the period encompassing how long it takes to make it. The raw process time is the actual time the work is being done. In the book *Lean Thinking*, by Womack and Jones, the authors use a cola can to explain that it takes 319 days (the cycle time) to do three hours of work (the raw process time) and that through the process, a.k.a. the value stream, eight firms and fourteen storage points are involved, the items are picked up and set down thirty times, and 24 percent of raw materials are scrapped.[16]

The problem long-lead items present is the need to schedule or forecast and the resulting matching problem (the challenge encountered in avoiding incorrect assembly when combining unique products with other unique products in an assembly operation).

---

16 James P. Womack and Daniel T. Jones, *Lean Thinking* (New York: Free Press, 2003).

he Matching Problem

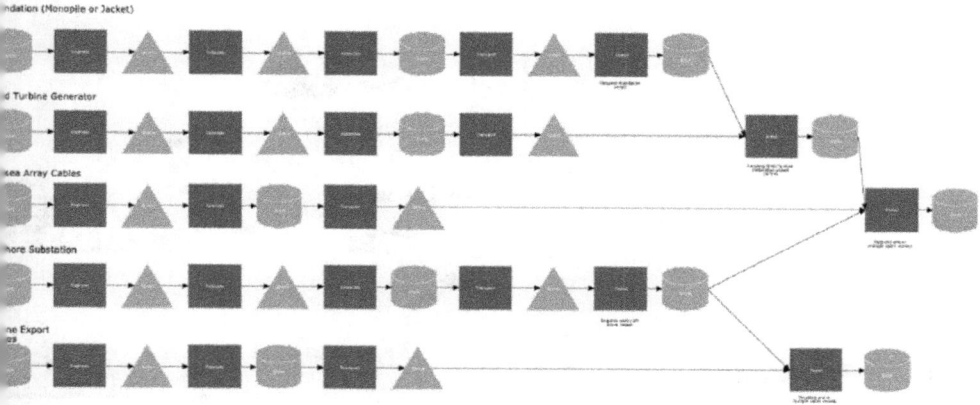

The matching problem relates to the challenges associated with bringing two or more items needed for a specific operation or task together at the same time. Contrary to conventional thinking that focuses on which items are late, we propose that items delivered before the final item are, in fact, early—or, in the world, of operations science, are WIP. This is exactly how measuring WIP enables us to understand and forecast time (schedule).

Matching Problem

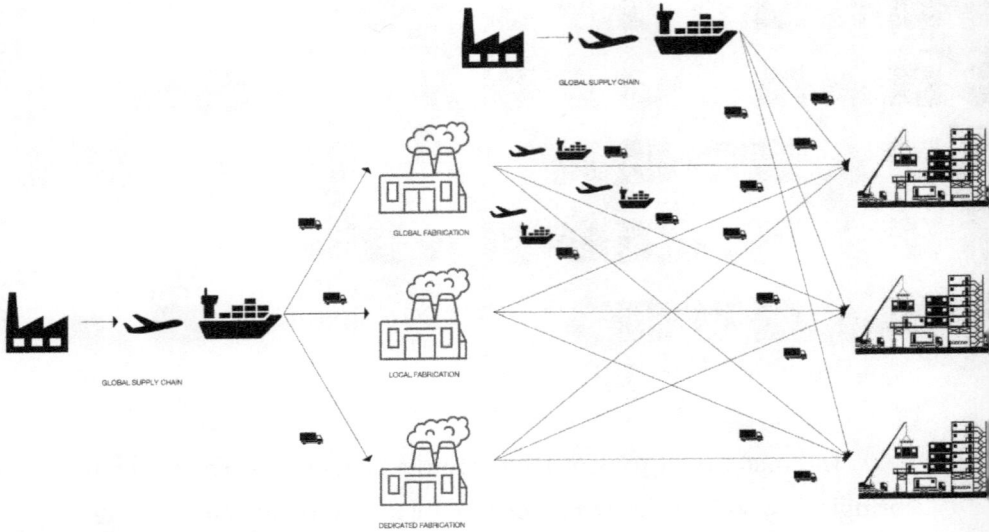

PMs don't care about materials waiting on people. There's nothing in the accounting system to account for this. Spending money equals making progress, so the thinking goes. By this logic, if you've blown 20 percent of the budget, you're 20 percent done, so it looks like you're moving along. This illogical system is reinforced by the fact that, superficially, it satisfies everyone: the owner, the construction manager, and the supplier (who gets paid).

For capital projects, lead times of twenty-four months are not uncommon. In the oil and gas sector, forty-eight-month lead times are not unheard of for equipment used in offshore field development and in the construction of large LNG plants.

This has become part and parcel of how the industry does things, and it's very much unlike manufacturing, which uses more of a just-in-time model.

Construction professionals are all too willing to agree to any lead time, no matter how inconvenient or counterproductive it is, rather than work to understand the cause of long lead time or reduce it. In practice, this means folks in engineering know that if they need three years to complete the project and it takes two years to get the material, they better figure out what they need and order it—when they should be saying, "Why the hell does it take two years just to get it? Can't we reduce that?"

As stated earlier in this book, the current percent complete accounting method used in construction masks the true cost of having unnecessary inventory during the delivery of a project.

This is the result of the cost associated with amassing, handling, holding, and preserving the materials (including the extra labor, equipment, and space) as well as the risk of theft (where one worker takes materials intended to be used in one area and uses it elsewhere) and obsolescence. These ancillary costs often exceed the costs incurred by not having the stuff there. Once again, we can use operations science to calculate the cost of an inventory strategy and to optimize the strategy, including selection, size, and location of stocks.

Everyone is asking the wrong questions. And if you ask the wrong questions, you're going to get the wrong answers.

THE CURRENT APPROACH TO MANAGING SUPPLY

Let's break down the cause of supply (mis)management and examine why the status quo is so faulty. As construction professionals, we are guilty of the following sins:

1. **Agreeing to long lead time rather than working to reduce lead time.**

   Long lead times force you to forecast and order materials so far in advance that sometimes you don't even know what you're going to need, which puts a squeeze on the time allowed for engineering. People tend to complain that "engineering is late," but really this just means "procurement might be early." This idea of the lead time has been accentuated by suppliers that have launched "lean transformation programs," so as they've become more lean, ironically, they have increased lead time so that they can be more efficient with their use of resources. The lead-time buffer allows them to optimize.

   During a visit to a plant owned by a major HVAC equipment manufacturer, we got to witness the most incredible lead-time story. In attendance were the president, VP of sales, plant manager, and several mechanical contractors. One of the contractors asked why the lead time was eighteen weeks. The first person to respond was the plant manager, who said, "Our lead time is two weeks." The president then explained to the plant manager that, in fact, the lead time was eighteen weeks. The plant manager reported he worked hard with his team to get the average fabrication and assembly cycle time down to less than a day (actually 7.2 hours) and that materials took less than a week from order to arrival. The VP of sales then responded that the lead time is eighteen weeks because "you contractors are always changing the order." The contractor then responded, "We change the order because the lead time is so long; we order anything to get into the queue, then change the order later as we get closer!" By this time, the plant manager was shaking his head in amazement.

2. **Allowing suppliers to work in the sequence and batch size they prefer rather than what is best for the project, all in an effort to reduce individual supplier cost.**

I saw this at a chemical plant serving the automotive industry. The plant issued a nine-thousand-line-item purchase order to the supplier of the pipe spools. The supplier used their MRP (materials requirement planning) software to run the production of the pipe spools in a sequence that was good for them but not conducive to site installation. The sequence of fabrication and delivery to the site needed to erect or install was not the sequence in which the work needed to be executed. It's like if you were building a house and buying stone for your kitchen or toilets for your bathroom—but you don't even have the rebar for the foundation. MRP is a software that tells you the best way to sequence the work based on the requirements of your demand. The more information you can feed into the software, the more effective it is. In essence, the software figures out how to optimize your raw materials use. The problem is that the more you put in, the longer the timeline extends.

The cost of the localized optimization is that the project suffers. Everything stacks up. Take, as an example, a large project for another chemical company we visited. They had their pipe spools fabricated and sitting somewhere five hundred miles away from the site. The steel was being installed, but no spools were going in. They were trying to determine why the project was taking so long. Cost reports looked good because they had spent a lot of money (in Era

2, spending equals progress). And they had done OK: all the pipe spools had been fabricated. But they had all been fabricated off-site as a huge inventory, without bringing them to the site.

A steel fabricator offers another cautionary tale. In their shop, the goal was to turn a big cutting machine on for structural steel members to run at 100 percent capacity. The welding bays could only take so many of those pieces coming in, so they built a little railroad to move this stuff around the yard. They were cutting in a sequence that was best for them but not for the project. They couldn't even handle the amount of material they were cutting, so they had to throw it on an internal rail system and move it out to the yard.

3. **Purchasing direct but taking a hands-off approach to managing suppliers.**

Basically, the folks in charge of purchasing say, "I'll issue a purchase order, so you as the supplier do what you gotta do as long as you meet our quality objectives." However, this hands-off approach is making things worse, not better.

We received a call from a refinery manager who was dealing with project delays—what did we find? Modules made in a sequence that was best for the supplier but not in the sequence needed by the site. Considering this was a major refinery and the lost revenue was significant, you would think they would offer some money to the supplier to work to the site installation sequence (versus the contractors' MRP system).

4. **Using inventory or more specifically stocks to buffer variability and attempt to increase productivity.**

   In the following example, we can see the common approach of using stocks of materials, in this case piles, in an attempt to protect the installation crew from running out of materials. The original strategy was to hold a stock that used $318,000 in cash with an 80 percent fill rate, while the reality was that inventory could be reduced 90 percent while the fill rate could be increased to 94 percent!

IDENTIFIER	DAYS PER PERIOD	MAXIMUM INVESTMENT	MINIMUM FILL RATE	TARGET FILL RATE	REORDERS PER PERIOD CASE 1	REORDERS PER PERIOD CASE 2	REORDERS PER PERIOD CASE 3
Completed Piles	7.00	500000.00	0.50	0.97	1.00	2.00	4.00

MINIMUM POSSIBLE FILL RATE	MAXIMUM POSSIBLE FILL RATE	MINIMUM POSSIBLE INVESTMENT
0	1.00	27539.26

Enter Optimal Target Values: Reorder Per Period [ ] and Target Fill Rate (%) [ ] [Calculate]

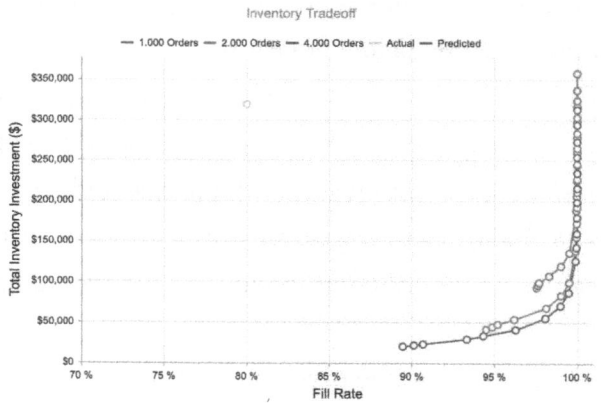

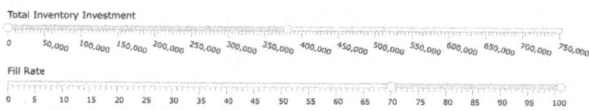

Inventory Tradeoff

— 1.000 Orders — 2.000 Orders — 4.000 Orders — Actual — Predicted

Total Inventory Investment

Fill Rate

Current Inventory Policy

| | | Current Conditions | | | | | | | | Predicted Results under Optimal Conditions | | | |
| | | Environment | | | | Order Timing | | | Order Quantity | | Replenishment Workload | Customer Service | | Investment |
Identifier	Unit Cost	Avg Dmd in Period (Units)	CV of LT Dmd	% LT Dmd Variability (Demand)	% LT Dmd Variability (Supply)	Reorder Point (Units)	Planned Lead Time (Days)	Safety Stock (Units)	Reorder Quantity (Units)	Days of Supply	Avg No. Reorders (Units)	Fill Rate (%)	Avg. Backorder Days (when short)	Value of Inventory
Current		165.00										80.00 %		$318,903.00
Predicted		165.00									4.16	94.31 %	0.08	$33,417.89
Completed Piles	$629.00	87.00	2.00	0.46 %	99.54 %	10.00	1.00		45.00	4.00	1.93	94.04 %	0.41	$17,508.20
Piles	$629.00	78.00	2.01	0.63 %	99.37 %	12.00	1.00		35.00	3.00	2.23	94.63 %	0.37	$15,909.70

5. **Attempting to forecast and/or schedule well in advance while operating in a dynamic, variable, and uncertain environment.**

This means that they believe they can predict what's going to happen months or years from now in a project, down to when a certain delivery is needed. And that's just not realistic, at least not under the current approach, which relies on predetermined critical path schedules to predict when materials will be needed at the point of installation. However, the ability to predict drops significantly beyond a day or two (as stated earlier in this book).

Example

A joint venture comprising internationally recognized contractors was spending more money and taking more time to do concrete work than the budget allowed. If the project continued at the pace of spend, the joint venture would suffer significant financial pain and loss of reputation. Specifically, the current rate of production would have resulted in being 55,000 hours over budget (as seen in the table below), but more importantly, liquidated damages due to delay would have been far more significant.

When asked to look at the problem, we began to see that deliveries of concrete were being provided late or canceled. This seemed strange, as the project had dedicated batch plants. Even more strange was the fact that a fair amount of concrete was provided by a third-party ready-mix supplier (which charged far more per cubic meter than if the project used their own batchers).

When mapping the current state of how concrete ordering, batching, and delivery processes were being controlled, it was determined that a classic approach based on issuing purchase orders followed by expeditors using spreadsheets to track the orders was in place. Furthermore, the "call-off" for concrete supply was based on predetermined dates set forth in the lookahead schedule (versus the actual conditions on-site).

To remedy the problem, a combination of production-based processes, control mechanisms and measurements were implemented. This included the use of standard work processes coupled with last responsible moment (LRM) scheduling, production plans for each shift, and the use of a pull signal from the workface back to the batch plant. The use of the pull control protocol enabled real-time synchronization between the site and the supply of concrete. Additionally, production-based performance measurements included reliability of workflow on-site, and reliability of concrete order quantity and timing, ratio of capacity used, etc.

The use of these measurements revealed that each of the work areas requesting concrete to be placed at 2:00 p.m. each day caused a surge in demand exceeding the capacity of the internal batchers. This created the need for additional concrete to be provided by the outside ready-mix supplier. An effort was made to stagger the demand over the day (reducing demand variability, a.k.a. heijunka, for those with lean background). The implementation of effective production control at both the site and for the supply of concrete enabled the joint venture to deliver ahead of schedule and under labor budget (see postimplementation results in the following table).

Operation	Pre-Implementation	Unit	Estimated Production Rate	Hours Above Estimate	Potential Additional Hours	Post-Implementa
Concrete Placing	9.28	Hrs/m3	2.00	+7.28	41,986	1.76
Reinforcement Fixing	17.00	Hrs/ton	12.50	+4.50	10,919	12.48
Formwork fix & Strike	5.22	Hrs/m2	4.00	+1.22	2,061	3.35

Courtesy of Strategic Project Solutions

THE PRODUCTION PERSPECTIVE

Instead of the project-management-centered or administrative-centered approach to supply/material management, we can adopt a production perspective. One area of significant opportunity is recognizing supply flows are in fact value streams (the end-to-end flow of work from production of basic materials, through processing, manufacture, fabrication, assembly, and delivery to the end user customer for a product or service—or in the case of a capital project, the on-site construction and start-up of an asset including all necessary design and engineering).

The Taylorist idea that the best, and fastest, way to do the job is by ramping up production to its maximum output, at every step in the sequence, is false. It actually takes longer that way. Now we can use modern OS-based tools to optimize these production systems, including setting appropriate inventory and capacity levels based on project needs rather than a wild-ass guess.

Example

The following real-world example depicts how a PSO study was done to determine the optimal inventory of materials on-site. Then a CONWIP-based production control solution was used to synchronize production and control the amount of WIP. The graph shows how this significantly reduced amount of inventory and allowed for far less use of cash than planned, while delivering a 99 percent service level (meaning 99 percent of the time, piles will be available when needed). This story plays on all day, every day, on construction projects throughout the world. Inventory and the associated cost and use of cash can be reduced with no impact in availability of materials, and in many cases effective production control allows for better availability of materials.

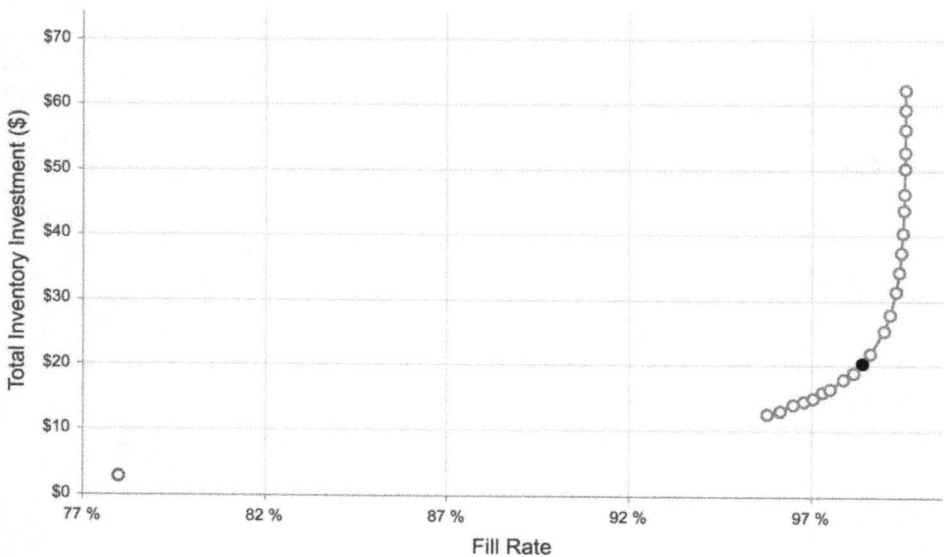

Because inventory space on-site is finite, the contractor needed to be aware of the necessary space requirements required for each component of the project. These space requirements are a derivative of the proper level of inventory needed to meet the demand of the construction teams.

The following inventory trade-off plot shows three curves that each represent different replenishment policies for both pencil and top piles. Each replenishment policy has a distinct effect on the total inventory investment (cash tied up in pile inventory) and fill rate (the percent of demand that is met from stock inventory) for the terminal piling production system.

The exponential shape of the three curves represents the trade-off behavior between total inventory investment and fill rate of precast piles. If a high fill rate is desired, the effect is more inventory on hand and, thus, more cash tied up in inventory. An optimal replenishment policy for the pile production system is one that allowed the contractor to carry as little inventory as possible while supplying enough piles to satisfy construction demand without interruptions.

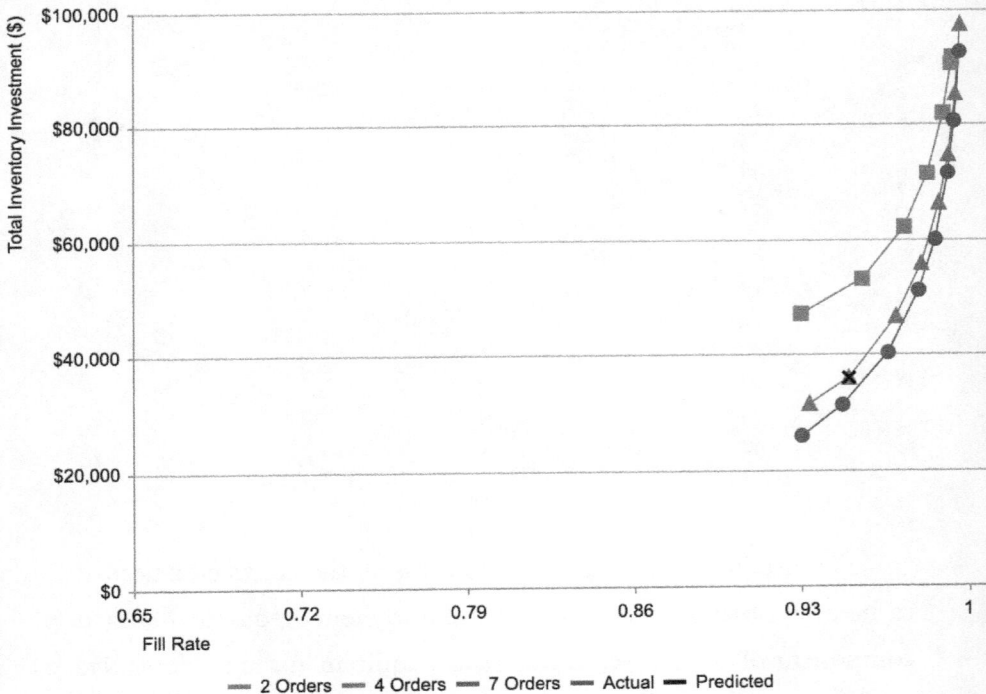

Inventory Trade-Off Plot—Piles

The square, triangle, and circle curves correspond to different order frequencies: two, four, and seven orders per week. It is clear from the plot that an order frequency of two orders per week will force the contractor to hold on to more inventory than the other order frequencies. Furthermore, the effect on total inventory investment and fill rate between four and seven orders per week are nearly identical. Therefore, an order frequency of four orders per week was chosen to reduce the volume of trucks flowing to and from site each week.

Based on the above analysis, the following replenishment policies were recommended.

M	AVERAGE ORDER FREQUENCY (orders/week)	REORDER POINT	REORDER QUANTITY	AVERAGE BACKORDER TIME (days)	REPLENISH-MENT TIME (days)
cil Piles	2	12	45	0.39	0.5
Piles	2	14	35	0.36	0.5

Piles Replenishment Policies

The above policies were chosen based on a 95 percent fill rate. This is an acceptable fill rate because of the short average backorder time, displayed above.

These policies establish the reorder points (ROP) for pencil and top piles to be twelve and fourteen, respectively, and the reorder quantities (ROQ) for pencil and top piles is forty-five and thirty-five, respectively.

When inventory position for each pile type reaches the ROP, an order shall be placed equivalent to the ROQ for each pile type. No orders will be placed for additional piles unless inventory position has reached the ROP.

AVERAGE CASH TIED UP OVER DURATION		
ORIGINAL POLICY	PROPOSED POLICY	OPTIMAL POLICY
$237,584	$68,753	$31,880

Piling—Cash Flow Benefits

The following charts demonstrate the benefits of optimal replenishment policy implementation.

Three cases illustrating the effect on WIP levels and cash flow throughout the duration of pile driving are demonstrated in each of the following charts: (1) solid line—original replenishment policy based on current state of decisions, (2) thin-dashed line—proposed replenishment policy given on-site demand at the time of this analysis, and (3) thicker-dashed line—optimal replenishment policy if designed into the production system at the time the purchase orders are issued.

Pencil Pile WIP Level with Demand

——— Original Policy ----- Proposed Policy —–– Optimal Policy

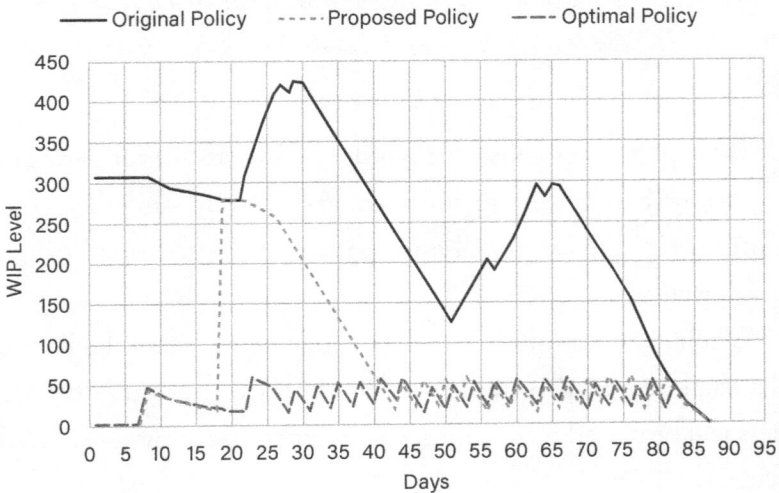

WIP Profile for Pencil Piles

Top Pile WIP Level with Demand

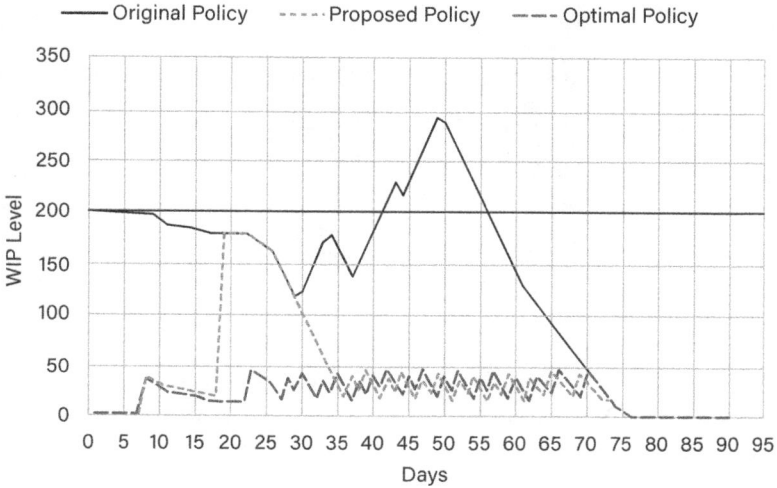

WIP Profile for Top Piles

Pencil Pile WIP Level with Demand

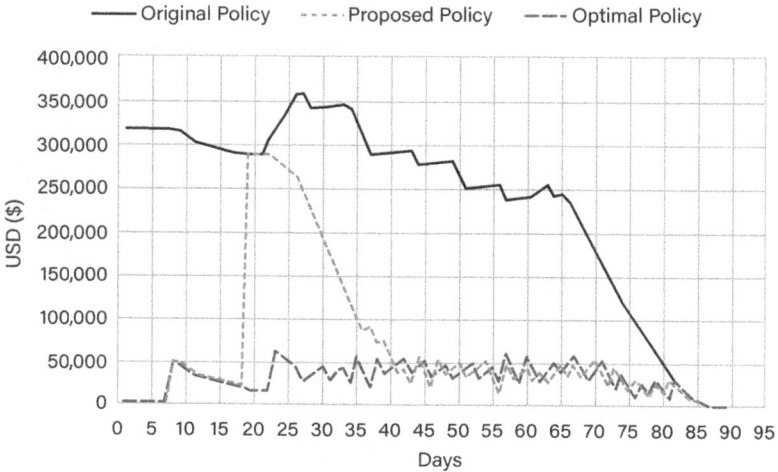

Cash Flow Profile for Precast Piles

The previous charts illustrate that effective replenishment policy allowed the contractor to better manage WIP and cash flow during a project.

Effective control of supply is required to implement optimal policies. The following schematic represents the basis of design of a supply flow control solution for piles.

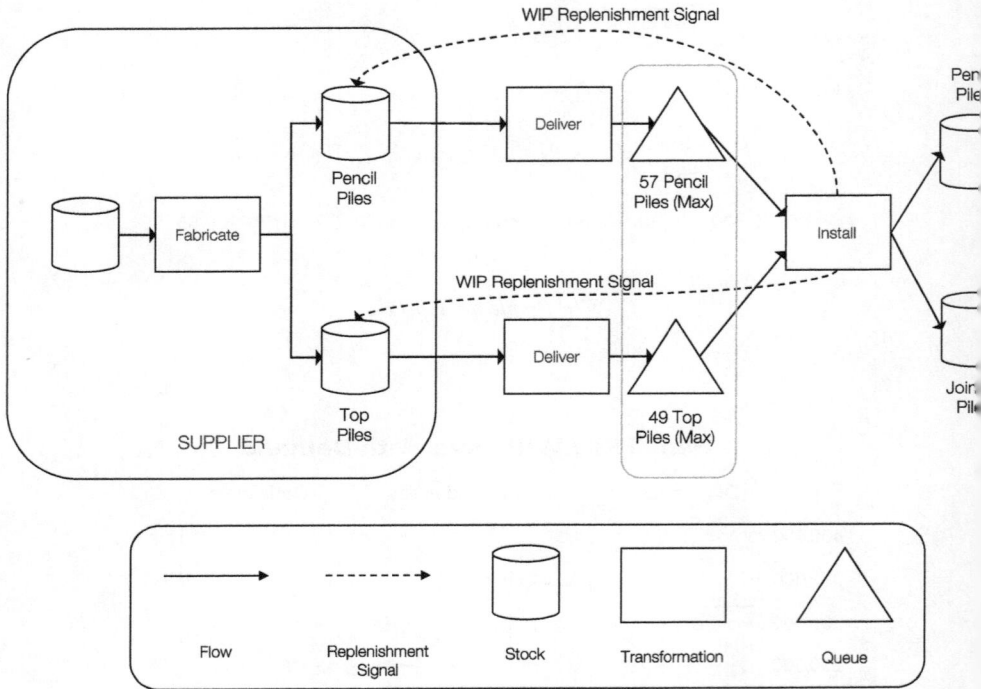

Supply Flow Control Design for Precast Piles

Unlike accepted practice that is focused on buying and expediting, the focus of the effort should be on production using the four verbs: (1) design, (2) make, (3) transport, and (4) build.

What, how, and where work will be done within a value stream determines whether an item is made to stock, configured/made to order, or engineered to order. As more work is pushed off-site into assembly facilities (e.g., assembling modules), integration and coordination become increasingly complex, pushing the ability of teams to effectively synchronize flows. (This results in the matching problem I mentioned earlier.)

Presently, what's happening in the trend of industrialized construction is that the decision has been made to move the work off-site. The thinking is that the more we do off-site, the better. For example, building a whole bathroom module or kitchen module, transporting it to the site, and installing it into the house being constructed. This is a big question in construction, with many pros and cons, and we'll address it later in the book.

It is also important to recognize that where and who performs the work in the supply chains has implications for downstream customers. Though a finished good delivered to the construction site may have been an engineered-to-order product as far as the site is concerned, other than the exact final location of the item, it is a made-to-stock item to them. It is delivered to site as a finished product; it is received and held until it is moved to the point of installation and installed. In other words, the site is merely pulling from an inventory of goods, be it aggregate or a one-of-a-kind-compressor.

SUPPLY CHAIN MANAGEMENT (NOT JUST PROCUREMENT)

It is important to handle different suppliers in different ways. You need to consider how the commercial agreement will influence lead time and sequence of production. Does the commercial agreement promote the desired behavior in the context of the production system goals?

The guy that's providing aggregate (a product supplier typically using a finished-goods stock to meet customer demand) needs to be handled differently than the guy whose patented technology you're licensing or the guy doing an engineered-to-order solution (a service provider), where you have to sit down and do some engineering, versus "Send us some aggregate so we can backfill a trench." Each have different business models and business drivers that are important to understand, as these business models and drivers will influence service levels, including lead time, sequence of production, and so on.

The following diagram sets forth a hierarchy of supply types, risks, and strategies: how to transact an associated contract type to ensure a productive relationship. Clearly sending a bid package to a company with a patent may not result in the response you desire, whereas creating an alliance with an aggregate quarry might not be the most efficient relationship (unless you are a heavy civil contractor that uses a mass amount of aggregate and you decide to own the quarry).

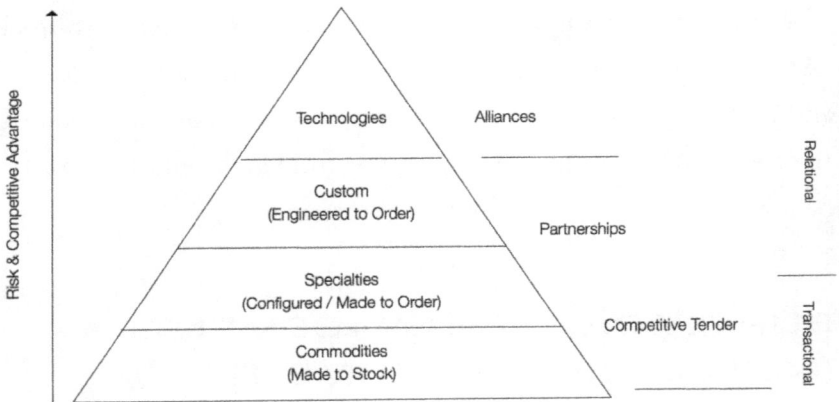

Ultimately, in looking at things through a production perspective, you can identify three levels of value that can be unlocked and captured from a supply chain:

1. **Eliminate unnecessary administrative work**: Work together better, and compress lead time. The elimination of unnecessary administrative work is beginning to be addressed by some companies, especially owners with advanced procurement systems. These owners understand there is no value in having suppliers submit an invoice on the suppliers' paper (that needs to be processed) when the supplier or contractor can submit a payment requisition in the procurement system.

> Ultimately, in looking at things through a production perspective, you can identify three levels of value that can be unlocked and captured from a supply chain.

2. **Optimize use of production resources across the value stream**: Who should do what, where? And, perhaps more importantly, let's not both hold stocks (you outbound and me inbound). For instance, a mechanical contractor may partner with a supplier that offers vendor-managed inventory, reducing the need to tie-up working capital wherein the supplier enjoys a captive customer.

Here's a little parable. You're a homeowner and you need new gutters on your house. The highly automated sheet metal shop comes out to your house, takes measurements, then goes back to the shop to prepare the estimate and sends it to you. You sign the agreement. They then inform the shop to make the stuff in accord with the specifications, then bring it out to site later and install it.

The Gutter King does things differently. He drives out to your house with a giant coil of aluminum at the ready. You agree on a price, and he starts custom-making your gutters then and there. The question is, Do we make the gutters in the shop or bring out a coil? It's a debate, and the answer is "It depends." (The solution is usually discovered in the production system modeling process.)

A real-life example comes from a study we did for a large energy company. They didn't know whether to make the pipes on-site or in a foreign country with much lower production and labor costs and then have them shipped. There were lots of opinions based on underlying mental models. We found that the on-site option was better, because the additional costs of handling, holding, and shipping offset the advantage in lower production costs. But that finding applied only to that particular situation. You must model it on a case-by-case basis.

For the most part, capital project production systems are an element of a standard value stream, as shown below. Raw materials are either extracted (mining) or recycled; these materials are processed into items such as steel, aluminum, glass, etc., which are then used to either manufacture, fabricate, assemble, or some combination of each. Understanding this structure forms the basis for understanding and optimizing project supply chains and production systems, which is done through various forms of modeling, including analytical and discrete event simulation.

3. **Strategic partnerships**: The highest form of supply chain management is to create alliances with suppliers that are important to your business. Make suppliers partners that enable you to drive revenue.

WHAT DO WE DO ABOUT SUPPLY?

The following are steps that allow us to understand, diagnose, and ultimately improve upon the issue of supply management through being able to offer differentiated solutions to customers.

1. **Recognize the true cost in time and use of cash associated with excessive inventory.**

   Inventory is not free. One example of this comes from the construction of a megaproject in the UK.

   One phase of the project required workers to make a bevy of rebar cages for cast-in-place pilings. In this case, the fab had a theoretical capacity of 12 cages per day, but the site could install only 10 per day. The cost of the fab per day is £41,760, and of the site, £16,600. The planners allowed 250 days for the installation of 2,240 cages, which was lower than the theoretical maximum of 2,500.

   However, the PMs had overshot their estimate of the theoretical capacity, which was greater than the actual capacity. The result, naturally, was a buildup of WIP that could lead to delays and increased cost of nearly £5 million (on a £14.6 million project).

   When we were asked to look at the situation, we discovered that since rebar cages were being produced at a faster rate than they were being installed, there would be a buildup of inventory. Well, you can't just stash extra rebar cages in a drawer somewhere. You need to store them—in this case, rented storage space, cranes, and labor for such large and numerous items would have added several million pounds to the total cost.

That wasn't even the worst of it. We discovered the initial estimate of ten cages per day was too high; in actuality, it was closer to seven, which extended project completion time and thus dramatically increased the cost.

The production system should have been designed with a limited buffer between fabrication and installation to account for variation in fab production. You can only allow so much WIP on the site. Thus, by limiting the storage to twenty units, we significantly reduced the total rehandling cost without extending the time for completion.

With that done, we focused on increasing capacity by evaluating all the steps involved in installation and optimizing the process at every stage, from receipt of the cage to its placement in the excavation to the final pour of concrete. Thus, we were able to achieve an average capacity of 9.1 cages per day.

Finally, we considered what previous project planners had not: the impact of cutting the rebar and delivering it to the site. These tasks had no bearing on capacity, but they did affect cycle time and WIP. The rebar contractor had the capacity for fifteen cages per day while the deliveries took one day and had ample capacity.

With these changes in place, our new model showed that this project could be completed on schedule. In fact, we beat that rosy estimate, finishing slightly early and knocking more than £4 million off the total cost compared to what had been forecast.

This narrowly avoided debacle demonstrates how production systems are often configured while remaining blind to unintended consequences. It also highlights how operations science plays a role in production throughput, inventory, and reliability.

2. **Understand the elements of lead time, including the actual cycle time and even raw process time for producing the item through modeling, and optimizing the supply chain including the design/engineering phase of the process.**

We often see cycle time to be a small fraction of overall lead time, with the root cause often being fairly simple to resolve.

When lead times are excessive or out of proportion to the cycle time, it's usually an indication that operations can be improved. If lead times are too vast, it's probably attributable to too much admin work, lengthy approval processes, or other internal bureaucratic snags, in addition to the desire to optimize. Fabricators look to optimize raw materials, labor, and equipment by working in large batches or batches larger than the site needs.

In one instance the owner issued bid packages for a set of engineered-to-order products to two suppliers, both of whom outsourced the majority of the production. Since it was a fixed-sum bid, and most of the work was outsourced, the two prime suppliers had to request bids from their suppliers. Over 20 percent of the lead time was dedicated to the bid process being undertaken by the potential prime suppliers.

3. **Put effective means of production control in place that enable control of WIP, versus attempting to schedule and having teams of expediters using spreadsheets to track materials.**

Using an Era 2–type schedule, which describes what needs to be done when, and calling up suppliers and saying, "I need the materials in advance, so let's get going!" project professionals track it on the spreadsheet while an expediter is calling around to track progress or embedding people into facilities that are producing supplies to ascertain what is actually going on. This is an imprecise and ineffective means of production control.

4. **Separate purchasing and commercial management from physical control of supply using blanket or other means of agreements. Consolidate purchases when possible.**

Conventionally, construction professionals want to control the purchase order rather than control the flow of work. In other words, if you're a supplier, I give you a purchase order, I track the results against the purchase order, and you tell me what sequence you want to deliver it.

It is a mistake to cede this responsibility to someone who, as competent as they may be, does not have a vested interest in, or even knowledge of, delivering materials in an optimal sequence (optimal, of course, for the project as a whole). Instead of just handing off purchase orders, we should be saying, "I need $10 million worth of X, Y, and Z, but I determine the sequence I want it delivered in."

With the former, you're just buying it; with the latter, you're separating the act of buying from the process of delivery.

5. **Align commercial arrangements with the type of product supplied.**

As stated previously, the construction industry, much like Western manufacturing in the 1970s, is blind to the true cost of inventory. This includes behaviors that result from commercial agreements. It is imperative that commercial agreements align to desired production system behavior. If left to their own devices, inventory will be used to make progress, generate cash flow, improve productivity, and so on.

ATTRIBUTES OF EFFECTIVE SUPPLY FLOW CONTROL

The right material in the right quantity at the right place and the right time is critical for achieving objectives related to construction cost, schedule, and use of cash. A supply flow control solution is the means to achieving this objective, which, while challenging, is by no means insurmountable. (The proof is in the pudding—my company has already been doing it for years.)

One must recognize that production systems, including value streams and supply chains, have built-in performance limits that cannot be improved without changing the design of the production system. Even so, production systems often do not approach their limits due to poor control. Additionally, most efforts miss the mark once again by concentrating on changing schedules (the demand of the production system) while ignoring the design of the production system, which is what dictates what will occur. A classic case of missing the forest for the trees.

The requirements for an effective supply flow control solution for capital projects include the following:

1. Quoted lead times do not drive decision-making. Instead, materials are fabricated, manufactured, assembled, and delivered when needed, but not so early that preservation related to storage (versus postinstallation) is required.

2. The use of resources, including inventory, labor, equipment, and space are optimized based on overall project objectives versus individual conditions (throughput and reliability of supply over optimization of a single operation within the process). We want to see the end-to-end work as best we can, within the context of the project. We don't want to make extra spools that we can't consume; we want to make them and test them as we need them.

3. The location and environment of all materials (by serial number or equivalent) is known at all times as they flow through the process of fabrication/manufacture, assembly, transportation, receiving, storage, and installation.

4. Limit the work at the point of final installation to final assembly or erection only; for instance, packaging and dunnage is removed at the marshaling facility/warehouse or not even used when not needed. I mentioned the story of our project at San Francisco International Airport. We were spending more time undoing the packaging than we were on installation. In contrast, the automotive sector makes use of dedicated steel or aluminum pallets that minimize dunnage and debris. In this, as in many things, we should take a cue from the auto sector.

5. Only what is needed when it is needed with no additional inventory is delivered and held at the point of installation. This includes removal of materials that will not be used in the short term). If you haul it out there and find you don't need it, there should be a means to take it back out as quickly as it arrived, instead of leaving it lying around taking up space and eating up cost.

6. It's agile and able to adjust to dynamic changes in demand from construction teams. This one is self-explanatory. Shit happens on-site. You must be flexible to accommodate unforeseen challenges and unexpected crises.

7. Loading, delivery, and off-loading capacity is used efficiently (including forklifts, cranes and associated rigging, trailers, and trucks). This refers to "parts presentation." For example, loading it onto the truck in a sequence that allows us to take it off efficiently.

8. Incorporate effective control of capacity and inventory instead of attempting to predict in advance what will be needed. The current practice is to schedule delivery of materials: "I need X by Y date." Unlike construction, in manufacturing, the guy at the end of the line dictates when he needs things. He sends a signal down to the front of the line, basically saying, "I'm ready; send me down the next thing." Essentially, everyone upstream to the install guy in the process is subordinate to him. This makes sense. This is known as a pull system—and almost no one in construction understands it, much less uses it.

> Unlike construction, in manufacturing, the guy at the end of the line dictates when he needs things.

9. Technology, such as software and IoT sensors, is used to automate key functions of the control system while providing important performance data. Again, rather than a bunch of humans with spreadsheets, we should have IoT sensors interconnected with GPS trackers on trucks, ships, planes,

containers, pallets, and on materials themselves—all of it "lit up" and feeding digitally into the production control system. Real-time total situational awareness.

10. Production-based performance indicators that allow effective improvement of the various production systems that feed materials to the point of installation.

CONCRETE RECOMMENDATIONS

With that said, what, specifically, can we do about the problem? The following recommendations lay out a workable, practical plan.

1. Design and engineer based on a design for life cycle framework, including implications of design decisions on manufacture, fabrication, transportation, installation, operation, maintenance, and decommissioning of the asset and its components. (We'll elaborate on this in the next chapter.)

2. Understand the configuration of each value stream and supply chain for all major elements being supplied to the project.

3. Locate suppliers as close as possible to the construction site instead of in a low-labor-cost location, especially when using off-site assembly strategies, including modules. (A caveat, à la the Gutter King analogy—there are exceptions to this rule. It is case dependent.)

4. Identify and remove any unnecessary use of capacity, inventory, and time contained in the project value streams/ supply chains using production system optimization. Do not

accept lead times as quoted; instead, map the supply chain and individual participant's production system to identify and compress lead and cycle time.

5. Minimize design issues while providing accurate bills of material through the use of a digital build. Chapter 8 will delve more deeply into design.

6. Control the flow of physical work through each value stream and supply chain using supply flow control (SFC).

7. Decouple commercial flows from physical flows such that purchase orders do not drive the physical flows.

8. Manage the bill of materials (BoM) such that the status of all materials flowing to the site and at the site is known. Do not use production packages or work packages to manage the BoM. The BoM shall be managed using the same unit of measure used for fabrication, manufacture, and installation. Deploy digital solutions including SFC- and IoT-sensor-based systems to control and monitor materials through supplier production to the site.

9. Define and design the on-site material-flow process taking into consideration the trade-off between inventory and capacity along with the implications of variability. Map, model, analyze, and optimize each material-flow production system using the production system optimization method-ology. Map, model, analyze, and optimize the production systems as early in the project delivery process as possible. For the most part, production processes are fairly standard, with decisions made mostly regarding where and who should do each operation in the process.

10. Use production packages as the unit of production that flows through the material-flow process. Limit the size of production packages to one shift for a crew to the extent possible.

11. Create production packages based on the last responsible moment date plus the agreed lead time for response.

12. Control the on-site material-flow production system (not the production packages) using the CONWIP control protocol. Link materials requirements (a production package) to the coinciding task in the production schedule. Use production-based performance measurements, including cycle time, throughput, utilization, and work in process, as the basis for optimizing/improving performance.

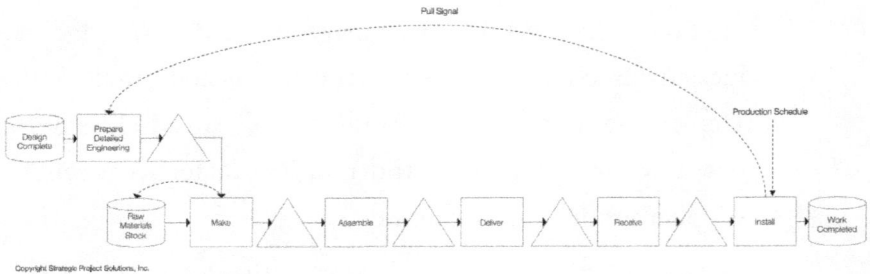

13. Do not schedule on-site logistics capacity (regardless of how well integrated it might be when it was created). Instead, execute work based on priority and timing specified in the production control system.

14. Forecast demand at the point of installation using production scheduling (not project controls forecasts and reports). Ideally, update production schedules in real time, but no less frequently than daily. To ensure reliability of the schedule,

include allocation of construction equipment, labor, and space in the production schedule.

15. Measure material-flow performance against demand through production-based indicators (versus conventional project controls indicators), and improve based on them.

16. Negotiate commercial incentives and legal agreements that align with desired behavior.

Through application of some of the elements outlined above, a colocation data-center provider was able to achieve the following performance improvements in less than eighteen months.

ASUREMENT	BEFORE	AFTER
nber of Process Steps	252	97
ivery Reliability	0-38%	98%
ation (PO to Install)	65-220 days	49 days[1]
S Cost	$800K/ Skid	$509K

rget 71 days

CONCLUSION

Ensuring that the right material in the right quantity gets to the right place at the right time is critical for achieving construction cost, schedule, and use-of-cash objectives. If materials needed to execute work are not available, labor costs increase, and schedules can be delayed. If too much material is on-site, cash is squandered or needlessly tied up while construction schedules may also be extended, to the detriment of the project. In both cases, unnecessary safety risk occurs, while quality may be jeopardized.

For the most part, control of materials to and on construction sites is done through purchasing and delivering based on a predetermined schedule or, more recently, attempting to use various work packaging methodologies. These approaches are not well suited for controlling materials. Construction sites are dynamic and mercurial, and the materials used in construction are often made or engineered to order. Not to mention, construction sites can be in remote locations or in congested city centers.

Effective control of material to and on construction sites requires the identification, definition, design, and control of the various production systems that feed materials to the construction site and ultimately the point of installation.

The next chapter illustrates this in greater depth by looking closely at the approach used during the construction of Heathrow Airport's Terminal 5.

CHAPTER 7

HEATHROW TERMINAL 5— IN A LEAGUE OF ITS OWN

THE MANAGEMENT OF SUPPLY for capital projects will benefit from recent developments in communications network capability and speed, computer power, and artificial intelligence.

In this era of intelligent production, we will no longer need to struggle with forecasting what will be needed where, by whom, and when, an arduous and inefficacious process dogged by expensive use of airfreight or fees for expedited shipping.

Building materials, hand tools, construction equipment, etc. will be instrumented. We'll be able to connect to production control systems to manage routing. We can control supply instead of trying (and failing) to forecast where and when we need things.

Construction professionals who have been involved in major capital works projects have seen it all—the good, the bad, and the ugly—as the following case studies illustrate. Era 3 supply management will help vanquish the ugly and make the good the status quo, not just the exception.

Heathrow Terminal 5 (T5) is well known not only in construction circles but also among the general public, simply because it was such a massive project and one that took twenty years to finish from conception to completion. In a project of this scale, you are going to find many cautionary tales as well as, as this case study will show, many small victories. SPS notched a few of these victories by playing a critical role in certain aspects of its development.

Led by the vision of Sir John Egan and an amazing team from BAA and Laing O'Rourke, an incredible feat was accomplished. I challenge anyone to come up with a project more innovative than what was done at T5. Concurrent design of product and process using advanced 3D and 4D modeling, standardization of design components, use of immersive reality, just-in-time delivery, project production control: it was all on display at T5, and it delivered results.

T5, the newest wing of the UK's biggest and busiest airport, was a staggeringly complex undertaking. Even in the best of circumstances, building an airport terminal is difficult. The T5 project presented additional challenges. Construction had to occur in a tight area, wedged between a high-traffic motorway, two very busy airport runways, and various other structures. There was no laydown, and a single point of ingress and egress allowed materials and vehicles in and out. Compounding the spatial limitation was a temporal one: deliveries of equipment and materials to and from the site were not permitted between 7:00 a.m. to 9:00 a.m. and 4:00 p.m. to 6:00 p.m., and it was not permitted to lay down materials for a significant length of time (a single day or even less). Moreover, most of the work was taking place after 9/11, which meant navigating a litany of new security restrictions.

These impositions would be tough for a project of any size, but the numbers give a sense of T5's scale. During its civil phase, each day the project consumed 5,000 tons of aggregate, 650 tons of Portland cement, 260 tons of fly ash, 200 tons of rebar, and over 100 tons of rebar cages.

SPS worked with BAA (the owner) and Laing O'Rourke (the main civil engineering contractor) to configure and deploy a project production system for all civil works and the T5 building phase. The integrated team created for this joint venture was dubbed the "Demand Fulfillment Team." We operated under the principle that the right information, the right materials, the right labor, and the right equipment in the right quantity would be delivered to the right place at the right time, every time.

That sounds simple enough. But in a project where, as a study commissioned by BAA showed, only half of what was supposed to be completed was actually completed on a weekly basis, achieving this was a tall order.

> We operated under the principle that the right information, the right materials, the right labor, and the right equipment in the right quantity would be delivered to the right place at the right time, every time.

At the same time, T5 was being delivered after several projects that ran well over schedule and budget (see the following table).

Project	Schedule (months late)	Over Budget
Wembley Stadium	8 months	$200 million
Scottish Parliament	48 months	$780 million
Bath Spa	60 months	$60 million
National Air Traffic Control	72 months	$550 million
New British Library	60 months	$3 billion
Channel Tunnel	12 months	$10 billion

APPROACH

To better understand the demand and potential behavior of the project production system, materials supply flows were classified using three categories:

1. Made to stock—Suppliers produce based on forecasted market demand for multiple customers.

2. Made to order—Suppliers produce standard products upon receipt of an order by specific customers.

3. Engineered to order—Suppliers produce unique products for a single customer upon completion of engineering (in some cases, suppliers develop detailed engineering for fabrication).

Specifically, the Demand Fulfillment Team found that the majority of commodity-based suppliers (e.g., aggregates, rebar, etc.)

made their goods to stock with large inventory buffers in their supply systems.

However, suppliers of engineered-to-order and made-to-order materials were unable to build large finished-goods inventories per market demand (because of the technical nature of the product). Therefore, these suppliers strove to convince the program's procurement team that long lead times were needed to ensure that such inventories would be available when they were needed.

With design and engineering less than complete, and the fact that even a program as large as Heathrow T5 results in a small order to a commodity supplier, the Demand Fulfillment Team focused its efforts on the high-risk engineered-to-order and made-to-order materials.

To govern this process, the Demand Fulfillment Team set forth specific rules for configuration of the project production system:

1. Use Consolidation Centers as control towers to coordinate all material flows into and from the site.

2. Realize "A Plan for Every Part" through forming assembly packages that equaled one day of work for a crew.

3. Deliver today what will be installed tomorrow.

4. Include in assembly packages everything needed to complete the work of the assembly package. Prevent accumulation of stocks on-site by returning the assembly packages that cannot be installed to the Consolidation Centers.

5. Use Consolidation Centers to form assembly packages, pulling from suppliers.

6. Distribute through marketplaces (site stores) personal protection equipment, hazardous materials, small-tool replacements, and consumables using "milk runs" and minimum/

maximum inventories (people could get gloves, wrenches, welding rods: whatever they needed, whenever they needed it).

7. Minimize lead times for fabricated products.

The strategy to implement these rules consisted of three main elements:

1. Use the extensive finished-goods inventories in the base or raw materials supply chains.

2. To the extent possible, mitigate detrimental variability resulting from rework, late deliveries, and so forth, and buffer remaining variability using capacity and raw material buffers (including locating fabrication and processing capacity at the site [e.g., reinforcing steel and concrete]).

3. Implement a constant work in process (CONWIP) control protocol to synchronize production from detailed engineering through to site operations to the extent possible.

Various initiatives were put in place to mitigate variability:

1. Physical resources including logistics centers, the aforementioned site store, and dedicated transportation equipment as control mechanisms to shield the site from variability by providing local inventory and fabrication, processing, and delivery capacity.

2. Digital prototyping supported by I-DEAS NX to model all components and assemblies that were needed on-site, including temporary works.

3. ProjectFlow (SPS|Production Manager) to enable implementation of the CONWIP production control protocol and

to reduce variability related to site through effective operations planning and control, including standard work and LRM-based scheduling.

The design of the logistics solution incorporated three different levels of control in the following order of priority: (1) physical, (2) software, and (3) human decision-making. Physical control mechanisms were used when possible (e.g., logistics centers, trailers, etc.), while software was the second alternative if physical means could not be used (e.g., ProjectFlow/SPS's Production Manager software for CONWIP control). The use of human decision-making as the means for control was limited to the extent possible.

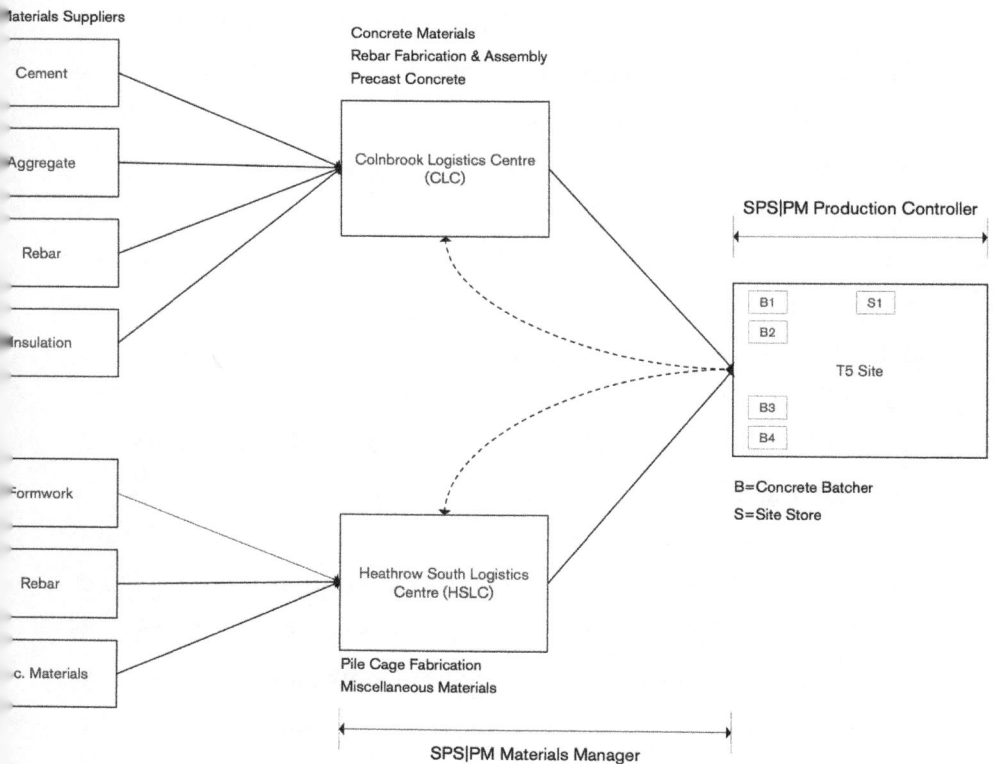

Materials Suppliers

Cement
Aggregate
Rebar
Insulation
Formwork
Rebar
c. Materials

Concrete Materials
Rebar Fabrication & Assembly
Precast Concrete

Colnbrook Logistics Centre (CLC)

Heathrow South Logistics Centre (HSLC)

Pile Cage Fabrication
Miscellaneous Materials

SPS|PM Production Controller

B1   S1
B2
T5 Site
B3
B4

B=Concrete Batcher
S=Site Store

SPS|PM Materials Manager

ht Strategic Project Solutions, Inc.

187

Two logistics centers were used to shield the site from variability (from outside the T5 program) and to allow "pulling" of materials to the site. Each center served a specific purpose. Rules for inventory buffers were established for deliveries into the logistics centers, to the site, and to the storage at the site. Typically, the site held today's and tomorrow's work, while the logistics centers held day three, and suppliers held day four and beyond.

T5 logistics centers and site stores served not only to shield site activities from supply variability but also to minimize the impact of the matching problem during the T5 civil phase.

The use of logistics centers, combined with the use of site stores, allowed for a reduction of process variability from factors such as traffic conditions, defective materials arriving to a congested site, and changes in plan. These centers, combined with the use of site stores, also shielded production from variability coming from the supply of consumables, small tools, and personal protective equipment.

Locations for nine site stores known as marketplaces were originally identified on-site to hold commodity type inventory, but because of space constraints on-site, the program was only able to deploy one marketplace. With this drastic reduction in inventory, from nine stores to one, the lead time for materials supply needed to be significantly reduced to increase the number of delivery cycles and maintain the desired service levels.

This was accomplished by establishing minimum/maximum inventory levels on-site, in the marketplace, and at the designated supplier, using kanban bins assigned to specific material types. A dedicated set of vans operated continuous milk runs, whereby they returned empty bins to selected suppliers and restocked bins to the site store and workface. This extremely reliable system enabled the T5 marketplace to replenish inventories up to three times a day.

In the end, we managed to reduce replenishment cycle times from twenty-four hours (next-day delivery) to two to three hours (same-day delivery). We reduced inventories from two weeks to one day of stock while increasing reliability of the supply system to 98 percent. Our efforts made it possible to supply more than eight hundred different selected made-to-stock products to customers on-site. We increased the number of dispatches from 100 to 1,200 daily in less than two months.

None of this could have been achieved without the help of cutting-edge technology. SPS's proprietary software, SPS Production Manager, was deployed to make this all possible. We iterated the software for the particular demands of T5, a configuration we called ProjectFlow. ProjectFlow was used by more than 1,300 active users to plan and control production (design and construction) as well as to synchronize supply flows with site demand based on the CONWIP control protocol. This allowed teams to do the following:

1. Manage site production on a daily basis.

2. Manage engineering production on a weekly basis.

3. Transparently forecast demand for specific materials across the supply network.

4. Synchronize across selected value streams, including detailed engineering, fabrication, and installation, using minimal inventory and time buffers.

While Production Manager/ProjectFlow was used mainly to mitigate process-based variability, digital prototyping via I-DEAS NX was implemented to mitigate product-based variability.

Concurrent digital engineering and CAPE were also used during the design process, including the concurrent design of the foundation, piles, pile cages, and pile-cage production system. During this effort, the team identified that the rebar manufacturing equipment and facility would need modifications. This happened numerous times during the project, including how best to erect the T5 control tower, build tunnel walls, etc.

REBAR DETAILING, FABRICATION, AND INSTALLATION

This example from T5 clearly depicts how the application of various PPM principles can be used to pinpoint opportunities to compress cycle time, including the role PPM principles play in identifying how other technologies, including 3D modeling, can optimize performance.

As the work on-site commenced, throughput of site production exceeded engineering production. Initial remedies to this problem included adding third-party engineering capacity. However, a quick exercise (by *quick* we mean one hour before lunch and one hour after) with stakeholders representing various operations within the rebar production system revealed the opportunity to reduce cycle time and WIP, and the need for capacity was sufficient enough to avoid adding engineering capacity.

In this instance, the value stream was well understood and well under control. Raw rebar was shipped into a T5-dedicated fabrication facility enabling control of all rebar activity other than the manufacturer of the rebar itself and transportation to the T5 fabrication facility. To minimize the impact of any variability associated with the manufacturing process and transportation, an inbound raw material inventory buffer was held at the fabrication facility using a minimum/maximum policy for the amount to be held at any given time. Standardization of various design elements was also used. This included "chairs" for underground cut-and-cover tunnels, structural walls, etc. With the front end of the value stream fairly robust and the design optimized, the next step was to optimize the on-site production system.

While looking for opportunities to compress the overall cycle time for the rebar production system, it was noted that a large percentage of the six-week lead time was the consequence of a complicated set of iterations between design, assembly, and engineering. The introduction of a rapid study, bringing together engineering, assembly, and fabrication stakeholders, coupled with the use of innovative digital technology for visualization of the fabricated parts and the steps required to install them using a 3D model of the construction process, eliminated most of the iterations responsible for the six-week lead time and culminated in a final process with a lead time of five days, shown in figure 5. To control WIP, an important feature was to install a CONWIP control mechanism to trigger the release of work at the detailed engineering stage.

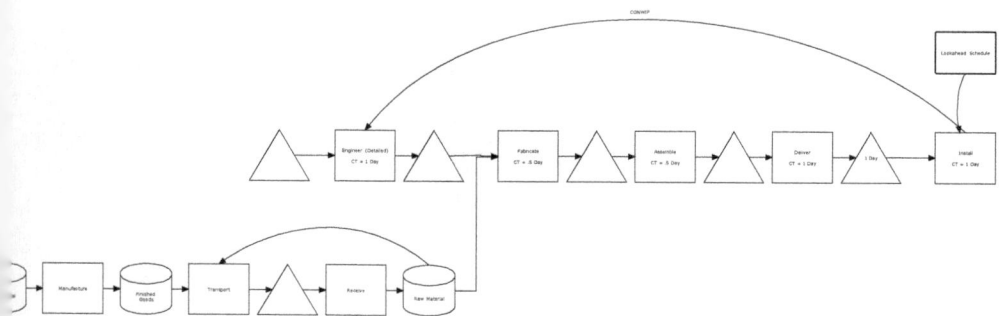

Heathrow T5 Rebar Production System and Production Control Approach

By fabricating parts only when requested rather than manufacturing to stock, not only were inventories reduced but, more importantly, the parts were ordered in the appropriate sequence for installation, avoiding significant variability that would have been encountered when searching for the correct parts in inventory.

The end results from applying PPM to the rebar production system, once identified within the Terminal 5 construction project, were substantial. The lead time for rebar drawing was reduced from six weeks to roughly five days. Most of the rebar drawings, up to 75 percent, could be detailed for preassembly. Designing for preassembly saved 20 percent on installation times. The rebar installation cycle times were reduced from two weeks to three days.

Results

In the end, it was reported that T5 achieved all milestones on budget.[17]

It's amazing what you can do when you question the old, established ways of doing things and bring in a new model. And T5 is just one success story of many that we've already achieved.

---

17   The House of Commons, "The opening of Heathrow Terminal 5," November 3, 2008, accessed September 18, 2023, https://publications.parliament.uk/pa/cm200708/cmselect/cmtran/543/543.pdf.

CHAPTER 8

DESIGN IS NEVER COMPLETE

WHEN IT COMES TO DESIGN, construction professionals and people involved in architecture or engineering are guided by a desire to "freeze the design." That means it's done. No more changes. The rationale for this is self-evident: finishing the design phase green-lights you for the next step and lends the project stability, certainty—something you can build upon. It's the nexus between concept and execution. But it just doesn't work this way. Design is never really finished. Freezing the design is a fallacy.

By the usual sequence, design completes requirements, engineering completes design, making things completes engineering, installing completes making things, and commissioning wraps it up. But when you start it up and begin to operate it—whatever "it" is—there are invariably changes that need to be made. Even the most thorough design phase completed by the most talented professionals does not bestow upon design an all-knowing, all-seeing, godlike power. Things get missed, unforeseen events transpire, new problems emerge, or better methods of how to do something come up. Sometimes, the design of one project can inform another project you are working on concurrently. And that know-how might be placed into a knowledge base to be used in other designs.

All these factors, which occur in every project, necessitate design modifications downstream from the actual design phase. Even the word *phase* is a misnomer because it implies a bounded period with a clear start point and end point. On the contrary, design is never complete. It continues throughout the life cycle of the asset. Even the decommissioning of an asset requires a design process to specify how to safely and efficiently do it, from disconnecting mechanical and electrical systems to the sequence for dismantling the structure, including the use of explosives for demolition.

WHAT IS DESIGN?

Design is a somewhat nebulous and context-dependent term, so let's define it, simply, as anything that takes the business or customer need and translates the need into a technical solution (these days most often represented through digital models, maps, diagrams, narrative, etc.). The solution may or may not be physical, but for the purpose of this book we will focus on physical—primarily capital assets.

The first step in the design process is to identify, define, and prioritize requirements. Requirements can be commercial, technical, or regulatory. For instance, the commercial requirement may be for the asset being considered to be able to produce a certain amount at a given cost to realize a specified profit, while technical may be related to life cycle and regulatory to what may be needed to meet environmental standards.

The identification, definition, or prioritization of requirements is a significant opportunity in the world of capital projects, but we spend less time on this compared to other industries such as aerospace or automotive.

Design encompasses conceptual design (are we constructing a ten-story building or a twenty-story one?); the engineering associated with realizing that design (science, mechanical, soil, structural, etc.); and finally, detailed engineering, which figures out exactly *what* and *how* we'll build. If structural engineering says the connections for the steel must be configured in a certain way, the detailed engineering will get into the nitty-gritty—the real, granular details of how that will be implemented on a technical level.

And finally, there is production engineering, which is mostly left to people in the field and considers the question "How will we execute the work?" For example, if we're installing something, do we use a crane? Where do we put the crane? How do we rig it? What do we do in the shop versus the field? Do we use a hand wrench or an impact wrench? And so forth.

There is no fixed boundary between *design, make,* and *install;* they are concurrent and bleed into each other. And as an industry, we continue to work to fast-track to the extent we can. Everyone throughout the delivery process is making design decisions, whether with respect to customer needs, scientific aspects of engineering, or detailed engineering in the field.

> There is no fixed boundary between *design, make,* and *install;* they are concurrent and bleed into each other.

From a cost, quality, safety, and time perspective, first- and second-order decisions are important. But all too often, due to architects and engineers not being involved in means and methods, third- and fourth-order decisions are made way too late in the process. The result is increased risk of safety incidents; reduced quality; and unnecessary cost, schedule duration, and use of cash.

ABILITY TO INFLUENCE DIMINISHES OVER TIME

Recall that the ability to influence erodes over time, as we can see in the figure below.

Later in the life cycle of the project, management has fewer options to intervene. Once you progress from identification, definition, and prioritization of requirements to the detailed design activities involved in actually making things, the cost of any given change spikes dramatically. Therefore, it is imperative to do as much as you can do in advance.

Some people take advantage of this dynamic by forcing design decisions early on, knowing that it will be difficult for anyone to modify it down the road. For example, an architect might impose the design they want by obtaining approvals from the relevant government agency. Well, now that the agency has issued planning permissions or permits for a particular design, as the project moves forward, you're basically stuck with it. It's going to be very difficult and involve a lot of intellectual labor, red tape, bureaucratic wrangling, waiting, and cost to alter it.

As illustrated by the following graph, as production engineering is pushed later in the project life cycle, the ability to influence declines; meanwhile, WIP is mounting.[18] To connect the ability to influence to OS, we incorporated the WIP curve indicating a WIP of decisions

---

18   This is a modified version of the graph from an article by Frederick W. Gluck and Richard N. Foster, "Managing Technological Change: A Box of Cigars for Brad," *Harvard Business Review*, September 1975, https://hbr.org/1975/09/managing-technological-change-a-box-of-cigars-for-brad. I have added the WIP curve.

being made, with the associated actions being taken resulting in less viable options, into Gluck and Foster's diagram.

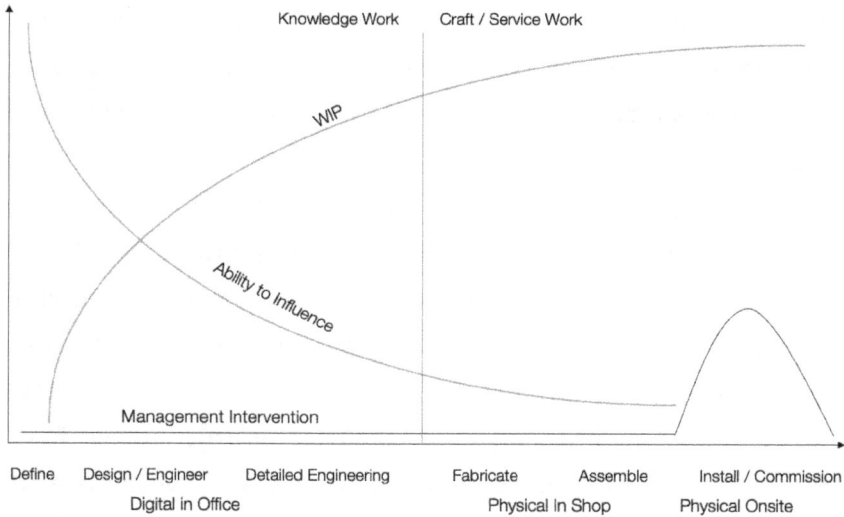

Ability to Influence Curve adapted from Gluck & Foster HBR September 1975

This is especially true of publicly funded projects or projects that involve a lot of regulations and government oversight. For example, a governmental agency or building department may want as much information as possible prior to construction (during the design phase), and in so doing end up forcing design decisions earlier rather than later, including decisions that you would preferably make down the road. The current approach, based on architects and engineers not being involved in means and methods, results in designs that do not take into consideration detailed production engineering efforts.

As the owner of the asset, you might not like it. But if you want the asset built, you have to play ball. The underlying problem here is that the industry functions on the paradigm that somebody is going to create various technical documents, drawings, specifications, and 3D models that provide a basis for other people to draw up estimates, proposals, and cost projections. You can trace the origin of this as far back as the nineteenth century, when municipal bodies sought to minimize graft by implementing open, competitive bidding or tendering procurement for public works projects and government contracts.

That system is still largely in place, especially for public works. To enable effective and efficient analysis of bids or proposals, a design must be put forth upon which contractors can prepare and submit a proposal or quotation, a.k.a. a bid. Once the low or winning bid is determined, a contract is awarded to the construction company, and, depending upon various conditions, the construction may have already begun or may be required by law to make agreements with the specialty contractors, or, if not required by law, they may "buy the job out" as they see fit.

Once the contract is awarded and the specialty contractors are hired, those specialty contractors begin to prepare and submit various technical data, which is sometimes known as submittals, or vendor data, describing the materials and methods they will use and how they will do the work.

You have a design team that's putting together a package, and that package might be a 100 percent complete design, or it might be a partial design, which is called *design-build bridging*. And that shifts the burden to the bidding team to complete that design. But either way, there's a package put together upon which to competitively bid the project.

We've examined a similar dynamic before in other facets of construction: it's another manifestation of the division of planning and doing. That's right: Babbage and Taylor. They're back! Design itself as fragmented into an array of specialists, from all manner of engineers (structural, mechanical, electrical, etc.), to architects to interior decorators to sound engineers and others. A lot of players crowding the field. Each has their specific tasks, but generally, their jobs are to provide documents to the downstream people handling delivery: the construction manager; the mechanical, electrical, sheet metal contractors; roofing contractors; paving contractors; and landscape contractors.

Over the years, design firms hamstrung by, among other things, the stringent mandates of their insurance companies and errors and omission policies, are precluded from doing anything related to means and methods. But you can't practically separate design from means and methods. They are, in many ways, two sides of the same coin. A design firm typically does not provide any input on how to perform the work or how the means and methods will be executed. It's problematic because what to do (the product) has bearing on how to do it (the process), and how things can be done has bearing on what is possible. In contrast, the auto industry does it differently. They've torn down the wall between design and production engineering, which is one reason why that industry is generally more efficacious.

As Babbage's specialization and Taylor's separation of planning and doing took hold, architects and engineers were unable to take advantage of the know-how possessed by construction companies. Rather contractors were forced to put this know-how to use by looking for opportunities to seek additional funds resulting from errors and omissions. As mistakes and omissions resulted in claims for additional compensation, law firms and consulting companies began to offer construction law and claims services. Architects and engineers are not to provide input or requirements related to means and methods or how the work shall be executed.

Fragmentation has also driven localized optimization supported by automation (i.e., software that is used to reduce steel tonnage in design and software to optimize sequence and batch during fabrication), increasing both complexity and the associated cost and time along with loss of revenue for the asset owner.

What sounds good locally is a disaster for the project and, more importantly, the business investing in the project.

DIFFERENT APPROACHES

While the advanced industries have sought to enable concurrent design and input from multiple stakeholders, the separation of design and construction became further engrained in how the construction industry operates.

This is also evident in how construction, as opposed to other advanced industries, has adapted new technology. Starting in the '70s and '80s, computer-aided drafting (CAD) and later 3D/4D modeling followed by product data management (PDM) and product life cycle management (PLM) were developed and used to improve upon antiquated processes. It was a big leap forward, which allowed these industries to efficiently and automatically do things that were not possible, or even conceivable, a few years before.

The development of CAD, 3D modeling, and PDM/PLM were adapted more effectively by blue-chip innovators in sectors like consumer electronics (Apple), automotive (Toyota, Ford), and aerospace and defense (Lockheed Martin, McDonnell Douglas) than in construction.

In construction, the first adopters of more advanced technology were specialty contractors, the guys doing detailed work. They benefited from computerized calculations of things such as material use or to make detailed isometric drawings that allow someone to install something in the field or make it in the shop. And the same to a certain extent was true for the specialty disciplines in engineering, whether it's structural, mechanical, or electrical. These advancements were particularly helpful for technically intricate, detailed work.

The use of 3D modeling has limited value to architects and CM firms because of the nature of their respective roles. An architect mostly occupies a conceptual (rather than detailed) role while the CM is more of a broker, winning contracts and hiring specialty contractors who handle the messy detailed and physical stuff.

The use of 3D/4D and PDM/PLM sometimes causes consternation because it was designed by the aerospace industry and then adopted by automotive industry and consumer electronics to allow for more concurrent and collaborative production within teams, teams that are formed during the earliest portion of the project, rather than teams that are assembled as the design progresses or is issued as a bid package.

These problems are not manifested in projects where the technology drives the need to make decisions that can't be made until the specialty contractor is on board.

The highlight of all this, for me, was when working as a specialty contractor on a large airport expansion project, the schedule was delayed over a year and the construction drawings were stamped "Not to scale." *How the fuck are you supposed to build this thing?* I thought. There has to be a better way.

The advanced industries do this in a totally different way. First, they have a completely different ratio of employees on payroll versus contractors. At one auto manufacturer I visited, I was surprised to learn that 80 percent of design professionals were in house. seventeen percent were contractors used to "flex," whenever more people were needed to support a project. And those 17 percent were individuals—humans: dedicated contractors, not companies—who would report to an employee on the payroll of the automotive company. The other 3 percent were outside people who could *not* be on the payroll because of various testing that would have to be done by a third party.

This is important because in construction, most of the people are *not* on the owner's payroll. Hiring has been divested or subcontracted out. In a given project, only 5 percent of the people involved are employees of the owner, and the rest are contractors, whether design firms, engineering firms, or specialty contractors. And contractors have a different (and often divergent) set of incentives and interests. This creates a problem: for every single project, you're creating new teams of new companies and new people. BAA had an interesting solution to this challenge. In the run-up to T5, BAA created a dedicated team of designers and contractors to deliver several projects and, in so doing, were able to learn to work together effectively. BAA went to great lengths to educate and integrate these teams.

Second, advanced industries place a greater emphasis on working *concurrently*. They look at concept design, engineering (including detailed engineering), and production engineering along with marketing and other elements *simultaneously*. What are we going to build, and how do we build it?

However, in construction, it's more piecemeal: looking at each element one at a time, from requirements to concept design to engineering to detailed engineering.

Because of that concurrent, as opposed to sequential, approach, those industries adapted the aforementioned technologies, such as 3D modeling or PLM, in a way that was more top down. When Lockheed and others developed 3D modeling in the '80s, they pushed that down onto their teams and suppliers. The advantage was that it allowed for *standardization* of tools: everyone using the same tools in pursuit of a common aim, because that's what the guys on the top had mandated.

The Lockheed Martins, McDonnell Douglases, and Dassaults of the word invented their own 3D-modeling software to support a way of working focused on concurrency and collaboration. Instead of working sequentially with the driving objective being to win a bid, they're working on all elements together, collaboratively, in a unified and coordinated manner.

In contrast, the construction industry is hobbled by fragmentation, because various specialty contractors were and are all using different software. That is the outcome of a bottom-up rather than top-down adoption of technology.

In effect, the software in use now, because it has evolved from the bottom up, has automated a process that doesn't work. That's the thing about software—it's a double-edged sword. Its capacity for automation is powerful and, if used right, advantageous. But if you automate the wrong thing, you just create additional problems. As Peter Drucker famously said, "There is nothing so useless as doing efficiently something that should not have been done at all."

The software has reinforced a way of working that is wrong and difficult to undo.

> That's the thing about software—it's a double-edged sword. Its capacity for automation is powerful and, if used right, advantageous. But if you automate the wrong thing, you just create additional problems.

FORGING A NEW PATH

Let's look at Toyota as an example. Unlike European and American companies, when it comes to design, Toyota adopted a couple of interesting strategies: they use (1) target costing and (2) a set-based design process.

Target Costing: A cost-management tool is used to set an anticipated cost for a product to be manufactured or services to be provided. It is a structured process that involves setting a target cost, analyzing the current costs of each component of the product or service, and then reducing those costs to meet the target cost.

It is a proactive approach to cost control rather than a reactive approach. Target costing involves a thorough understanding of the customer's requirement and the design process and manufacturing processes so that cost savings can be identified and achieved. It is used to remain competitive and to maximize the profitability of a product or service.

Glenn Ballard at UC Berkeley has been leading the way in the development of target costing that he terms *target value design* for construction projects.

Set-Based Design: Toyota designs multiple solutions for a problem or multiple options for a situation and *holds those options open longer*. They'd resist freezing the design and wait as late as possible—until the last responsible moment.

Set-based design is a design methodology that focuses on the design of a system or product as a whole rather than its individual components. Set-based design is based on the idea of creating a set of design solutions that meet the design requirements instead of focusing on optimizing a single solution. This allows for flexibility and creativity in the design process and can help reduce costs and time to market. Set-based design also allows for rapid prototyping and testing of design ideas. It can be used in all aspects of product design, from conceptual design to detailed engineering.

For example, Toyota might design five different power train options and keep carrying them on as viable options, avoiding locking themselves into any one thing. That allows them to see the total landscape of how everything fits together at the strategic level.

In the Western world, design was based on freezing and then building upon the freezes—a completely inflexible approach that might make sense in theory, if you never have to unfreeze anything. But that's just not viable in complex systems of production. The rework caused by undoing a freeze was a vexing problem which Toyota cleverly found a way to evade with its LRM approach.

As a result, Toyota was able to lower its design costs and improve design quality because they avoid or minimize the freeze-induced burden of rework. They could create more options at lower cost, faster.

In construction, there is huge pressure to freeze as soon as possible, namely because that's when you can start pricing it. In manufacturing, it's $profit = price - cost$. Toyota can only charge so much for a car before consumers refuse to buy it. In construction, because it's a service model, $price = cost + profit$.

In advanced industries, and in companies like Toyota, everyone is sitting around working together, and most of these collaborators are on the payroll.

Toyota also tends to give 80 percent of work to one supplier and 20 percent to another. They tell the number-two guy, "You can be number one, but you have to come up with better ideas than the other party." That incentivizes both parties to continually improve upon their technical design. Concurrency and collaboration, all digitally enabled: this is why Toyota remains a model in design for companies worldwide, including those in industries beyond the auto sector.

Product life cycle management (PLM) systems, pioneered by Dassault, SDRC (now Siemens), Parametric Technologies (now PTC), and other firms, represent the next exciting phase of design technology, and hopefully construction will take a page out of their book. In PLM systems, complete configurations of a product can be made. In practice, that means "We can produce configured-to-order vehicles, such that for the most part no Land Rover that comes off the line is identical to any other."

They're taking the idea of a platform design and allow for a radical degree of customization, configurations, or scenarios—an exponential number of them.

With advanced PLM systems you can develop configurators that give the customer the ability to basically design their own product. The construction industry has lagged behind, but some forward-thinking players are adapting. For example, Goldbeck, a German firm, excels at this. They are a business to watch. They do all the design and most of the fabrication in house and also do site construction and have configured products. They have shown what's possible via a more advanced approach to design using configure to order based on a platform design.

Call it *mass customization production*—that term is paradoxical, but it has been adopted by some of the leading global companies that have figured out how to take the best of both worlds from high-volume production and specialization/customization.

The automotive and aerospace industries offer a stark contrast to capital project design and construction practices. As market conditions became more complex and dynamic, new ways of design and engineering were needed. This resulted in the development of concurrent design and engineering, a work methodology emphasizing the parallelization of tasks (i.e., performing tasks concurrently), which is sometimes called simultaneous engineering or integrated product development (IPD).

RAMETER	ADVANCED INDUSTRIES	CONSTRUCTION
ocess Flow	Concurrent	Sequential
oduction Engineering	Done early in process	Done late in process
of Materials Accuracy and Precision	High	Low
st Management	Cost as input to design	Design as input to cost
ganizational Structure	Team based	Discipline based
tio of design done in-house	High	Low
pplier involvement in early phase	High	Low

Today automotive companies and other advanced industry players are leveraging the metaverse in an interconnected virtual world that exists in the form of a computer-simulated reality. This is made up of virtual environments, such as virtual worlds, augmented reality platforms, and mixed-reality applications, with the aim to better understand customer desired value, behavioral trends, do design and engineering, and communicate work instruction, among several other use cases.

RECOMMENDATIONS TO IMPROVE DESIGN: WHAT SHOULD WE DO?

1. Manage (Identify, Define, and Prioritize) Requirements

The industry is not good at identifying, defining, and prioritizing requirements. People like to solve problems, and they enjoy coming up with solutions, but the reality is that perhaps the most important part in the project process is the definition of requirements—when you are "translating" the voice of the customer to the voice of the engineer. The customer usually frames things in terms of the object of what they need rather than the utility of what they need. When they say, "I need a new office building for my company," what they really mean is, "I need a structure that can house my workforce." It's kind of like if someone says, "I need to lose weight"—what they really mean is "I want to get healthier" or "I want to look better." It's incumbent upon the people who are working on the requirements to understand why that thing is needed, because therein lies the value.

As Peter Drucker stated, "And what the customer buys and considers value is never a product. It is always utility, that is, what a product or service does for him."

I recall one project where a prospective client of a design-build firm, a beverage company, wanted to expand their brewery because they wanted to ramp up production. The production engineer hired to assess the needs of the project concluded that the beverage company didn't need to expand their facility; instead, they needed to optimize how they were doing their production with the brewery that they already had. The design-build firm lost the project, but they gained a customer for life because they didn't build something that wasn't needed.

I once heard an architect say, "We must push the envelope to define the boundaries." As exciting as this may sound, it does come with a cost and a risk. For the majority of projects, it is far better to "map the design space" or, simply put, to define the allowable options based on various requirements.

Requirements fall into one of three categories:

Physical: even the most visionary minds are limited by the laws of our physical reality (i.e., "you want a twenty-story structure, but if you build more than ten, it'll collapse").

Stakeholder: there are numerous stakeholders in any project, starting with planning and building government agencies you need to get approval from, environmental impact studies, building codes, lenders with their own stipulations, sureties (people who bond the project and have *their* stipulations), insurance companies of various stripes, the local community, customers, suppliers, etc.

Business: obviously, there are various parties who are seeking to earn a profit and need to build assets to do so. Economics, including cost to build, cost to operate, and revenue, are key elements of business requirements.

Contrary to popular belief, business requirements are often the areas of greatest flexibility, even though most firms place them on the top of the priority list. Most enterprises operate under the assumption that business needs are the most important, when in fact stakeholder needs (not to mention the immutable laws of physics) take precedence. Moreover, these three different types of needs are often at odds.

To minimize risk, large corporations have gone to great lengths to implement stage gate project processes. These processes consist of predetermined decision points (typically accompanied by presentations, meetings, etc.) to decide whether a project should proceed to the next phase, be held at the current phase for additional analysis, placed on hold, or even terminated. It is not uncommon for final investment decisions to be made long after the project has invested far beyond what is needed to make a go-no-go decision. Each stage gate requires you to fulfill some criteria or render a decision to move forward.

The reason for this is the fact that the requirements of planning authorities and environmental agencies and the desire to understand the financial viability of the project drives the need to do work far in advance of what is needed to actually construct the asset.

Owners don't want to get too far down the road or get ahead of themselves under the stage gate process because that would mean wasting money. In other words, they want to avoid expending a lot of up-front resources only to eventually pull the plug on the whole project altogether. But the demands of stakeholders—for example, an environmental impact study or the restrictions of a municipal planning commission—might require so much design and planning that it puts the owner out of whack with their preferred stage gate process. We call this "being a little bit pregnant."

For example, I mentioned a government agency and its regulations for building a hospital. To obtain approval, your technical documents must be so detailed that the conventional project delivery process based on bidding becomes infeasible, because the design must be so complete that the specialty contractors need to be involved much earlier in the process than they are used to. It's hard to garner the involvement, much less the commitment, of contractors for a project that is, for all intents and purposes, just a gleam in the all-knowing, all-seeing eye of government health regulatory bodies. But this government agency doesn't care about your stage gate process. They have their own mandates and interests to which they are beholden.

In other words, the stage gate doesn't work in practice because it ignores stakeholders. It's another form of project administration created by people enamored with bureaucracy while ignoring the real-life demands of the environmental agency, the planning commission, etc.

2. Constrain the Design Space

Step back and ask: What does all this mean?

Rather than learn about all this later, let's put it all on the table to see what we can or cannot do. Instead of pushing the envelope to define the boundaries, let's understand the requirements of the planning commission, let's understand the requirements of the business, let's understand the building codes, let's understand what's going on in the market. Let's put all that together, further define those requirements, and then prioritize them. Now, our decisions are not going to be as broad as they were before, but they will be decisions that we can build upon.

> In other words, the stage gate doesn't work in practice because it ignores stakeholders.

So instead of the envelope-pushing, boundary-defining architect saying, "The planning commission says we can only have a ten-story building, but let's go for a twenty," he should say, "We're really going to get a ten, so let's design to a ten, but maybe we can go down four floors and get the fourteen." These requirements need to be understood and not brushed aside.

At a more granular level, detailed engineering benefits from having libraries of specified parts and materials to choose from, along with policies related to tolerances, use of parts, etc. that limit unnecessary exploration in items that do not meet requirements.

3. Involve Specialists Early to Enable Concurrency

There is no reason not to have specialty contractors at the table early in design, as is done in the auto sector. They bring valuable know-how related to detailed production engineering that others don't have. In particular, specialty contractors do more projects than design firms and construction management firms, so they have more in-the-trenches insight. As a matter of fact, the separation of planning and doing requires specialists to be involved early.

This is easier said than done. The challenge is twofold: (1) owners and construction management firms are eager to know the cost of the design, while (2) specialists have been conditioned for generations to put in a bid on something, so when you invite them to the table, they just say, "Tell me what you want, and I'll give you a price." They don't know how to collaborate once invited to the party since they've never had the opportunity. This causes tension between the owner, the design team, and even the CM team, who are scared to death they might not get the best price. But the best use of your investment comes from tapping their know-how to figure out the best way to do something—since they're the ones who in fact do it every day. And doing so within the context of clearly defined and understood target value, including the total cost. And to do so early in the design process, where detailed engineering and production engineering can influence decisions related to product design, including how best to make, deliver, install, and maintain the systems and parts needed to build and operate the asset.

4. Organize Based on Systems (not Disciplines)

Instead of organizing by discipline, organize by system. For example, there would be a foundation team that includes all specialists who work on the foundation (soil engineers, concrete contractor, steel contractor, etc.).

The means and methods issue results in separating the product design from the process design. Architects and engineers form one side of the hierarchy, while construction managers and the specialty contractors inhabit the other side. Companies go to great lengths to integrate and create high-performance teams that are divided by the project structure and the point in time when each member joins the team.

Following is the organizational structure for a new headquarters for a company in the UK that was built in 2004. The idea was to form teams based on systems in which each team had all the players needed to represent design, engineering, fabrication, logistics, construction, commissioning, and operation.

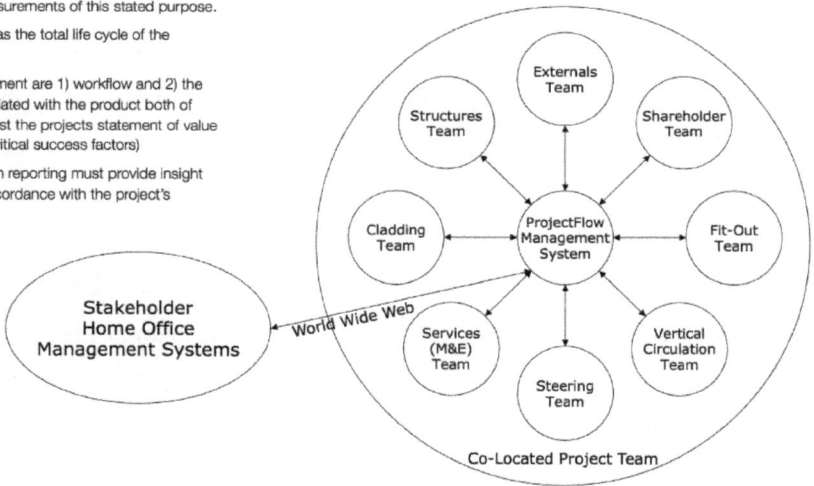

Theirs is also a movement out there to colocate everybody in one space—everyone working in the "Big Room." But the problem with colocation depends on the workflow or the process through the design process; you may have people that don't have any work to do (remember the resource-based schedule earlier in the book). So that's the trade-off for colocating: a loss of capacity and the associated value. And of course, there was the learning of new ways of working remotely during the recent COVID-19 pandemic.

213

5. Map and Control the Process

Projects are unique. The output is a custom product, typically brought into being by a team put together for that particular project, whose constituent members represent various entities. And the actual process of the design execution is poorly understood. There is little time spent on mapping the design process (i.e., "here's what needs to happen, and how").

There's a lot of "doing" going on. The structural engineer knows what to do. The mechanical engineer knows what to do. The soils engineer knows what to do. Everybody knows what to do. They have the know-how and the skills and the computer systems to do it. But the integration of the flow of work is not understood globally. We need to map that design-process flow so that everyone involved understands the flow of work and thus can control it.

We also want to home in on places where we need to decouple. Sometimes different players are in a kind of standoff, like a showdown in old Western films, where two gunslingers face off in the long-shadowed evening of some dusty desert town—only this time, no one wants to draw first. Design is not sequential. It's iterative. Managing that flow is critical, and understanding who goes first and *how* they go first is also critical. What we observe is either no means of control or attempts at using bar charts or even CPM for managing design. During the past several years, there has been increased adoption of methods and tools for determining and agreeing to the flow of design work. But these tools lack the ability to address the issue Fondahl mentioned, that resources will influence the time it takes to execute the work. Though the flow of work is important, we must not forget the importance of effectively managing capacity and WIP. And WIP is critical, as we want to avoid rework caused by making decisions too early or without the proper input.

WHAT ABOUT AGILE SOFTWARE DEVELOPMENT?

Based on reported success in their IT Departments, executives are looking at more widespread application of agile in their companies. Specifically, the adoption of agile software development methods to the design of capital assets.

This is an interesting development. First off, we must recognize that agile processes (iterations, sprints, scrum, stand-ups, etc.) and team-based organizational structures have been used in construction since the early 1900s. We must also acknowledge that software development projects and capital projects are fundamentally different.

Construction of physical assets requires a complex supply network with lots of engineered-to-order parts. Work is both technical and physical. Outside stakeholders such as licensing (both professional and commercial) and trade unions have influence over who does what work. For the most part, this is not the case for software development. Regulatory requirements result in large-batch or waterfall-type project delivery. Though self-imposed governance processes result in unintended consequences, construction, for the foreseeable future, cannot abandon a waterfall model. Unlike software development and numerous product development efforts, construction incorporates significant regulatory approvals. These regulatory approvals and the associated filings include environmental impact studies, planning authority/commission approvals, building permits, and certificates of occupancy, among just a few examples. Regulatory agencies require specific processes to be followed and information considered for each of these approvals. Additionally, it is common for governmental and other public agencies to require competitive bidding and tendering for the process of selecting engineering and construction contractors.

To facilitate this process, instructions to bidders or bid packages must be assembled and distributed.

For the most part, these regulatory requirements do not apply to software development. Though at some point, as with all things, governments may choose to regulate software development. One can see privacy and security being two primary areas of focus. How would this regulation be implemented, and what challenges will it impose on the software development process? If this occurs, perhaps software development will need to look to construction for solutions.

Construction is related to the physical world. Therefore, chemistry, physics, and the associated fields of engineering and materials science apply. Foundations need to be built prior to structures being erected; concrete must cure before achieving its structural integrity.

The physics of construction establishes constraints to product design; the design process; and decisions related to capacity, inventory, and variability. Changing the sequence of work requires a trade-off. For instance, decoupling a foundation from the structure such that the foundation design does not delay the remainder of design can be done by "overdesigning" the foundation (accept loads that may not ever be placed on the structure). However, this trade-off comes at additional cost to the project.

Interestingly, the major tech companies actually have used the waterfall method of design for their consumer products.

Agile in itself is a questionable practice because it assumes we have unlimited capacity. It believes that there always will be capacity to do work; namely, because it's technical work, that there'll be knowledge workers ready to go. And that's not always the case.

Digital technology has revolutionized every facet of life and business, and there's no reason why construction can't harness its untapped power either. But applying technological solutions must be done in the context of a clear-cut, level-headed strategic vision, not just throwing things at the wall and seeing what sticks. In the last third of the book, I'll talk about an autonomous, digital future, address how we can tie all these elements we've been discussing together, and propose a framework for transitioning confidently and adroitly beyond the ineffectual status quo.

CHAPTER 9

A NEW CHALLENGE— DEPLOYMENTS

CONSTRUCTION IS A FIELD that drives changes in society and at the same time is driven by broader trends in that society. For example, the invention of reinforced concrete in the mid-nineteenth-century enabled us to build things bigger, stronger, and faster, from drains to hydroelectric dams to highways to cathedrals (the Sagrada Família Cathedral in Barcelona, for example—incidentally, a project that is still unfinished after about 140 years. Talk about schedule overruns.) This innovation allowed civilization to grow and advance at a rate that was previously impractical.

The dynamic works in reverse too: construction responds to broader trends and must adapt to meet changing needs and evolving technologies. Presently, construction is responding to the sweeping changes, such as digitalization, green energy, and small-batch production, that are revolutionizing industries old and new with dizzying speed, from energy to manufacturing to telecommunications to pharmaceuticals.

These technological trends also shape how our industry delivers capital projects. It's causing confusion and shaking up long-standing ways of doing things. We construction professionals must remain on the avant-garde of the trend.

As a simple example, look at health care. In the past we'd build a single large hospital. But with the rise of distributed health care, where diagnosis, testing, and treatment take place in several more localized (or even virtual) settings, as an A&E or construction company, you mandate for what to design and build changes accordingly.

The keyword guiding how we engage with these changes is *deployment*.

The old challenges are being replaced with new. The many-to-one supply chain problem—having to buy all the stuff you need and ship to one site—does not apply anymore. Now we have the same complexity in the supply chain but many sites to which to deliver materials. The game has changed: instead of doing one big project, how do we manage one thousand smaller deployments?

The old challenges are being replaced with new.

To get the world to net-zero carbon by 2050, BloombergNEF, in its "New Energy Outlook 2022," estimates required expenditure of over $5.5 trillion per year up to 2030, increasing to $7.4 trillion in the 2040s. Similarly, the IEA estimates $100 trillion plus will be required over the next three decades to reach net-zero carbon emissions by 2050. This does not include the massive investment needed for digital infrastructure, so it is imperative we learn how to effectively manage deployments.

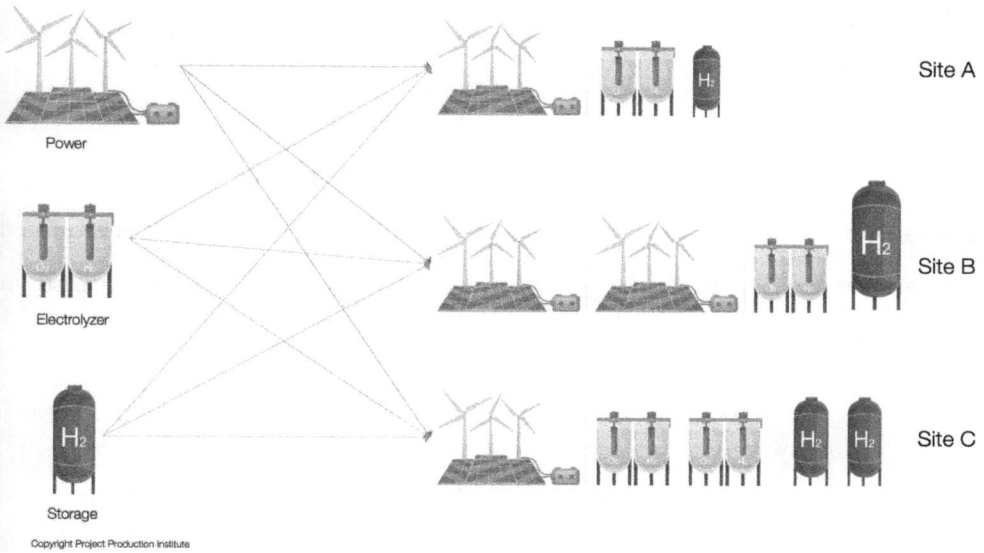

Copyright Project Production Institute

The reason why deployments are more difficult is threefold:

1. You have to get a lot more permits: whereas before one or two permits for a big job would suffice, now you can expect to file for dozens, hundreds, sometimes even thousands. As a matter of fact, a very large telecommunications company explained that securing permits is their biggest challenge.

2. You have more complex site geography: there are many sites in many locations that you have to make work with your supply logistics.

3. You have to rely on local contractors and their labor to do the work. And there may be many different contractors locally. You may have to work with local A&E firms because they have the relationship with the building departments or possess the local licensing to work in that jurisdiction.

CONSTRUCTION	DEPLOYMENT
One location / jurisdiction	Multiple locations / Jurisdictions
Dedicated design firm	Numerous design firms
One permit	Numerous permits
Engineered to order	Typically, standard product that is localized
Many suppliers feeding to one location	Many suppliers feeding to numerous locatic
Single lead contractor	Multiple contractors

So now you need a broader, more diffuse team of people because of this geographic dispersion. It has financial and accounting implications as well, because of the disparity in labor costs in different locales. A bricklayer in San Francisco costs more than in Des Moines.

A NEW PARADIGM: CONFIGURE TO ORDER

The old, laborious, and probably soon-to-be-obsolete method is to design for each site. The more advanced companies are saying, "We can come up with a base design of the asset that covers 60–70 percent and then localize it to complete the rest, meaning adapting to the demands of the local site."

One of the challenges of localization is navigating permitting requirements and local codes. I know of one company using configure to order that ran into a problem when they tried to integrate their base design with the local building codes. Everything was fine, with one exception: the company used the color red to denote positive conductors, and the municipality where they were building wanted the conductors to be purple.

One big question is determining at what point does one do configured to order versus engineered to order. Where is the midpoint between them? Engineered to order commences with designing the thing. Configure to order has a base design upon which to select options. For instance, though a car company may have a platform design, the options are 10—virtually infinite.

We also have to learn how to pay more attention to site conditions. When doing deployments, you face new levels of complexity when engaging with the geography of the numerous sites. Questions abound. For example, if you're building a hydrogen refueling station, you must consider what the conditions of the soil are, whether there is an existing gas/petrol tank, whether you can just drop in a new technology or whether there is a combustible situation, whether there is water and power available, and how the storage tank will be refilled.

As an analogy, think about the distinction between software and hardware. In software, you can basically do what your creativity, innovation, the capacity of the programming language, and the hardware allow. With hardware, you have to think about the physical parameters, as well as the requirements of other parties, such as UL Solutions (formerly Underwriters Laboratories), and so on.

But construction of an asset not only incorporates hardware and software but also takes it all to another level, including the sheer size of the product being produced, the site conditions, soil conditions, etc. Many involved in the software industry and even hardware or consumer products struggle to understand the requirements and complexity of construction. Recently someone explained

> Construction of an asset not only incorporates hardware and software but also takes it all to another level.

that we have spent thirty years focused on software and during that time we forgot how to make and build things.

HOW DO YOU CONFIGURE THE PRODUCTION SYSTEM AND THE SUPPLY CHAIN FOR MAKING THIS STUFF?

The question to ask is, "What are we going to buy versus make?" Then, if we're buying it, from whom? If we're making it, how and where?

At one deployment of an advanced technology, the company was struggling with whether to build a megaplant or several smaller plants located in customer facilities. They had some unique security concerns since they were utilizing cutting-edge technology, the type that drew the unwanted interest of foreign intelligence agents. With a single plant they can monitor who goes in and out of the facility. If there are multiple facilities or localized sites at the customer's location, then security is far more challenging. They had to conduct a supply chain cost-benefit trade-off: weighing the advantages of trucking raw materials in and finished goods out of a centralized megaplant versus placing multiple units close to their customers or even inside their customer's facilities, as is common.

Another example includes an energy company that traditionally would build a single megafacility. They have learned that it is better to build multiple facilities on an as-needed basis. The reduced batch size/WIP gets the company to revenue faster with less cash outlay. But building several smaller facilities increases site and supply chain complexity. The company was even trying to understand how many in what size at what time—an unnecessary question if the old one megafacility approach was used.

Owners must also consider their relationship to contractors. If a project involves multiple contractors spread over a geographically expansive area, the contractors doing the physical work (mechanical, electrical, structural) may not be able to travel across the deployment's geography. So how do we ensure the best value of supply chain when there are multiple contractors?

Once again, let's turn to the automotive sector for some inspiration. Advancement in 3D modeling and project life cycle management techniques can teach us a lot about engineered to order versus configured to order.

For example, the auto sector makes ample use of configurators that allow you to take your base design and rapidly configure it without having to start over and create a whole new design. Automotive design tools utilize parametric modeling (intelligent models wherein when one thing changes, all related items get automatically updated) *and* configuration-based design. A platform is configured based on customer needs and wants (brown leather interior, sport seats, etc.) but using a standard platform upon which to build.

Finally, operations science and its tools related to modeling and control provide the means for solving the supply chain and multisite challenges associated with technology deployments. OS assists us in determining what to make, what to buy, and how much of it, respectively.

CASE STUDY: BETTER PLACE

Better Place was a Palo Alto–based start-up launched in 2007. They sold battery-charging and battery-switch solutions for electric cars using a mobile-phone-type business model—their mission was to electrify vehicles around the world and alleviate one of the long-standing problems that deterred people from switching from gas-powered vehicles to electric. The company grew quickly and raised a significant amount of venture capital. But multiple flaws in their business model (along with a reported out-of-control spending on marketing) ultimately sunk them.

Nevertheless, the vision driving Better Place's deployment was fascinating. SPS was hired to help figure out some of the thornier questions surrounding deployment of this technology—specifically, challenges such as how to take the installation cost of a single charge spot (including the associated infrastructure which ranged from $2,300 to $500 per unit). Now, most construction people would see this as a very small project. And if each site gets ten charge spots, then it is a small project. But that is the hidden risk of deployments. The program consisted of potentially millions of locations globally. In the game of deployments, small wins result in massive benefits.

They really had several things going on from a deployment perspective. First, they were designing new technology for charging electrical vehicles *and* for swapping the batteries within electric vehicles.

Second, Better Place was looking to partner with auto companies on the design of electric vehicles powered by one battery in the car, with another battery that could be charged separately or swapped. Battery swap became one of their product offerings and one of the innovative features that made Better Place an overnight success (albeit a fleeting one). The idea was that you would pull up at a station, drop out your battery, and put another one in within a few minutes, all without getting out of your car. They were designing the swappable batteries, the cars that could accommodate them, and the infrastructure to enable the swap.

Third, they had to figure out how to deploy the infrastructure to enable this to happen. How do we find and evaluate sites where we can install charge infrastructure or battery-swap infrastructure? And if we swap them out, where does the old one go, and how do we charge it? What does that infrastructure look like? Where do we get the energy for the charging station? One of their ideas was to operate dedicated, large-scale charging "warehouses" on a vast scale. It never came to pass, but it was certainly novel. Imagine what that kind of infrastructure might look like if millions of consumers had bought into this system.

Fourth, how do you integrate and coordinate the design of the technology, the design of the car that complements the technology, and the deployment of the infrastructure so it all comes together (in other words, resolve the matching problem)? That's the kind of multifaceted, multidimensional conundrum you have to untangle when it comes to deployment—there are many moving parts, each a puzzle unto itself.

Let's take just the deployment. Do you put the charge spot in in advance, or should you only deploy the actual infrastructure (the conduit and the transformers and the cabling or wires) to later install the charge spots? The charge-spot design was originally integrated with the electrical infrastructure. When we worked with Better Place, one of our contributions was, using the five levers, to decouple the charge-spot design from the electrical circuit infrastructure so that they could install the infrastructure, then at a later date install the charge-spot. The charge spot could be put on and "clipped in." This changed the LRM date for the delivery of the charge spots (allowing for more innovation) while the infrastructure was being installed.

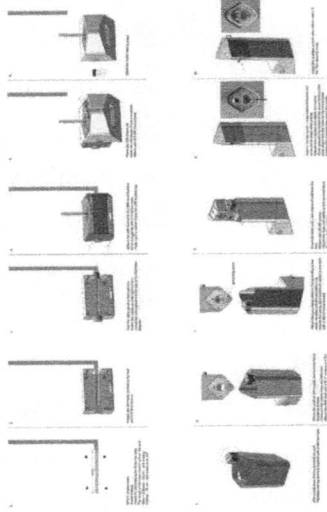

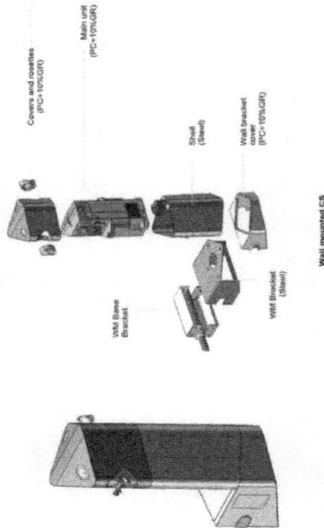

Base Design Charge Spot Installation Process—Note Integration of Electrical Junction Box with Bracket

Base Design Charge Spot on Left and New Version Charge Spot on Right
(Note Decoupling of Junction Box)

And then at an even later date, say, when the next generation of the charge spot was rolled out, the old charge spot could be removed and replaced rapidly without affecting the existing infrastructure.

Meanwhile, on the process side, we introduced the idea of how to use augmented reality, laser scanning and constraint-based automated engineering tools to take an existing site and have the computer figure out the design. Drawing on the concept of configure to order, we let the computer figure it out based on the number of charge spots, loads, sizing, and routing of the infrastructure needed for these charge spots. The charge spots could be mounted on a wall in a parking structure or along the street or sidewalk as a pole-like structure.

Then—again using the five levers—we analyzed how much capacity and WIP we would have at any given time to support this program of the charge-spot design and manufacturer and the infrastructure deployment to receive a charge spot, based on the stock coming out.

At this point we were able to go from ten days and four hundred hours of work to one day and forty hours of work! This won't be surprising if you get your head around fundamental project production management and apply it.

CONCLUSION

The big idea is that how things are done in most major industries is changing, resulting in smaller, multisite projects with multiple stakeholders, and more of them. That is very different from managing a big project, for a variety of reasons.

This introduces more complex supply chain problems and forces us to reckon with tough questions about the geography of deployment and how to handle labor needs when work is performed over a diffuse and decentralized terrain.

In light of the movement from big-and-centralized to smaller-and-decentralized construction, given the role it plays in building the infrastructure for society to function, it must also adapt. Green energy and the turn away from fossil fuels, digitalization and the internet, and small-batch production are reshaping human life and economic activity. For example, instead of one big plant making a large number of the same thing, let's make custom products in smaller, localized facilities. That's a mode of production that will become predominant in the twenty-first century.

Accordingly, the ratio of big single-construction projects to smaller-scale deployments is going to decline. Until now, the majority of construction has been bigger single projects, but through energy transition and digitalization we expect things to turn toward more of a deployment model.

In the next chapter, we'll take a closer look at the future of construction—how the emergent technologies of today will reshape the construction industry of tomorrow.

CHAPTER 10

THE FUTURE: AUTONOMOUS, INDUSTRIALIZED, AND DIGITAL

IMAGINE IT'S 2030. You're looking at a site where a new hospital is being built. Autonomous, self-driving vehicles are delivering materials (tracked with GPS sensors connected to the internet) while robots off-load them. Other robots lay brick, weld pipes, and put up wallboard. Every physical thing on-site is tagged with a sensor for data gathering about location and environment, from the point of the fabrication to the assembly shop to when it's transported on the road, rail, or ocean, until the point it is installed. As are the machines doing the work and the tools being used by humans—smart tools. Meanwhile, off-site automated factories are producing supplies with barely any human intervention. And machine learning enabled constraint-based design systems are leveraged to design the buildings and the infrastructure themselves.

This futuristic scene is visionary, but it's not science fiction. Already, machine learning, IoT sensors, cutting-edge data analytics, warp-speed information networks, and robotics are being deployed to make construction faster, safer, more precise, more efficient, and generally better. Right now, the tech is used in a piecemeal fashion (a little of this here, a little of that there), and much of it is in its infancy. But when it all comes together, it will give us the ability to deploy what we term Intelligent Production[19] or self-forming/self-controlling production systems.

ADAPT OR DIE

The technologies of the future are coming whether construction is ready or not. Inventors and engineers, for the most part, aren't getting out of bed excited to invent a robot, improve processing speed, develop quantum computing (someday), or improve machine learning techniques just so they can fix the construction industry. But that doesn't mean we can't and shouldn't integrate this tech into our own work. Indeed, we *must*, or we'll be left behind.

> **The technologies of the future are coming whether construction is ready or not.**

So how will the construction industry benefit from this tech? How can we leverage it? And how do we ensure we don't use it to automate or digitize what should *not* be automated or digitized? As Bill Gates said, "Automation applied to an efficient operation will magnify the efficiency ... Automation applied to an inefficient operation will magnify the inefficiency."

---

19   Intelligent Production is a registered trademark of Strategic Project Solutions Inc.

Why is it important? ⟶ Business Problem or Opportunity
         ↓   ↓
How to approach it? ⟶   Conceptual Framework
         ↓   ↓
What to go do?   ⟶   Digital Solution

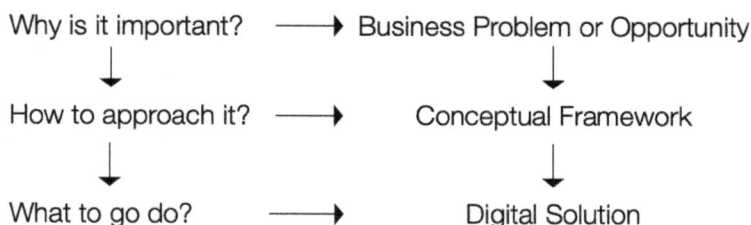

In response, PPI has established the following graphic to assist professionals in understanding how emerging autonomous, digital, and industrialized production all fit together and in so doing modernize construction. This graphic sets forth how various emerging technologies build upon the foundation of operations science (OS), enabling the realization of modern construction. Why is OS the foundation? Depending on who you ask, the purpose of a project is to create an asset, service, result, outcome, deliverable, etc. With the keyword being *create* and the synonym for *create* being *produce*, we need to focus on the creation or production aspect of a project. And, as stated in this book, OS provides the basis for understanding and influencing production—therefore it is the foundation.

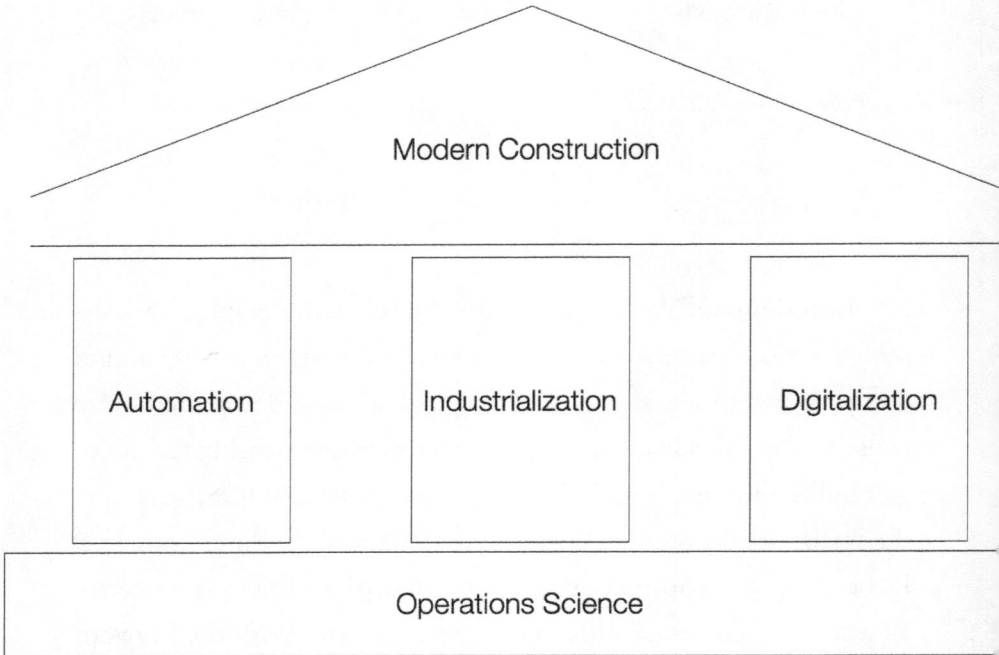

AUTONOMOUS AND ROBOTIC

Certain companies are advancing quickly in an area that not long ago felt like a contrivance dreamed up by Hollywood. Hilti builds semiautonomous robotic drilling systems. Kodiak Robotics already has self-driving long-haul trucks on the road. Human safety professionals are also along for the ride, but eventually, those vehicles will be 100 percent computer operated. Peterbilt is also getting into the autonomous vehicle game with its Peterbilt 579, and its sister company Kenworth has the T680. And Peterbilt is not some bullshit start-up. They're a household name that's been around for nearly a hundred years.

Caterpillar has been investing in autonomous development for decades in support of the mining industry and offers telematics in support of several other products. Soon excavators, loaders, demolition robots, and other types of robotics will automate manual labor and repetitive, difficult, or dangerous tasks while boosting efficiency. For example, Boston Dynamics's bipedal robot, Atlas, is now being shown doing various tasks on mock-ups of construction sites.

So this is real tech, not some pie-in-the-sky start-up smoke and mirrors to seduce venture capital firms looking for the next hot new thing into forking over bags of cash.

INDUSTRIALIZED

Significant investment is currently being made to industrialize construction, with major investors such as Warren Buffett's Berkshire Hathaway entering the space. As outlined earlier in this book, industrialized construction is best understood as the application of high-volume manufacturing concepts, methods, and tools to the design and construction of buildings, industrial facilities, and civil infrastructure.

Key elements of industrialized construction include standard product designs at the part, subassembly level but with the ability to create a custom final product, the application of advanced PDM/PLM systems to enable constraint-based design and configuration, and moving fabrication and assembly work from a construction site to a controlled environment.

Practitioners of industrialized construction see the majority of work, including fabrication and assembly, using flow-based production processes and standard products that can be configured to order based on customer and other stakeholder requirements. Like in advanced industries, PDM/PLM systems are used for design, engineering and for provision of numerically controlled code to automated production tooling.

No doubt, as all this unfolds, the metaverse will continue to play a bigger and bigger role in support of customer engagement during sales and design, factory layout and optimization, CAPE, and work instruction.

Practitioners of industrialized construction see the majority of work, including fabrication and assembly, using flow-based production processes and standard products.

With their implementation of each of the technologies outlined above, one can argue that Katerra was the leader in this area.

Regardless of how and what happens in this area, massive amounts of data will be produced by the PDM/PLM systems used to design and configure and from the equipment and tooling used to make, transport, and assemble.

DIGITALIZED

You might not associate a literal brick-and-mortar industry like construction, so rooted as it is in physical labor, heavy machinery, and materials, with digitalization. But autonomous and digital technology has reshaped all facets of society, and construction is not impervious to it.

This trend opens up numerous exciting possibilities for applying data collection and analytics to a wide array of processes and functions. For example, now companies like Milwaukee support Bluetooth tracking, allowing for precise, real-time, systematic transmission of data about how those tools are used, where, by whom, etc. Companies like Apple and Google provide means for tracking people as well and generating data about their movements through smartphones. RFID, Wi-Fi, satellite, cellular: they're all means of transferring data.

The graphic below illustrates how these different functions and data capture and transfer methods interact. As you can see, the majority of the technology is in place and in use, almost all of it in your mobile phone!

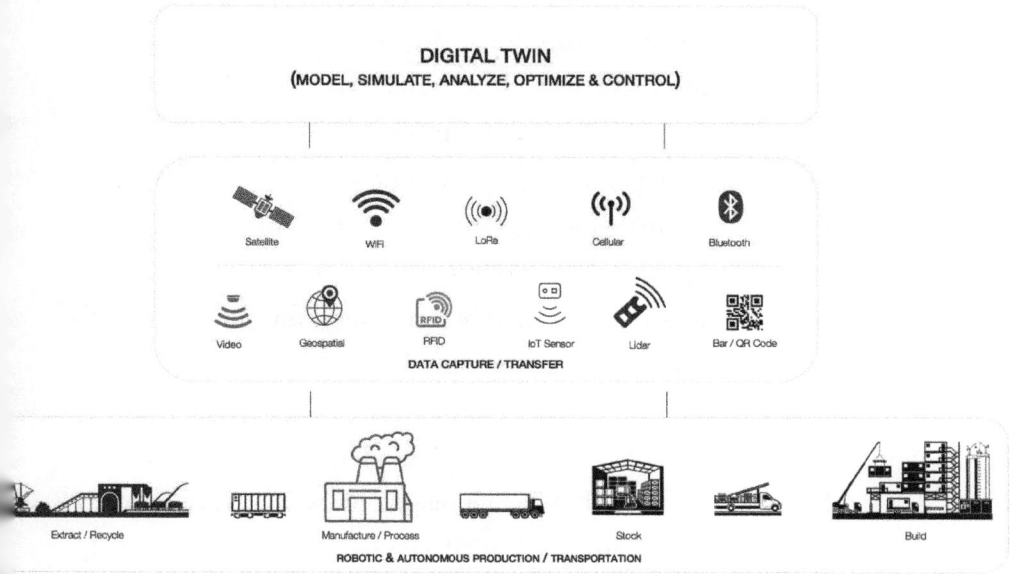

DIGITAL TWIN
(MODEL, SIMULATE, ANALYZE, OPTIMIZE & CONTROL)

Satellite WiFi LoRa Cellular Bluetooth

Video Geospatial RFID IoT Sensor Lidar Bar / QR Code

DATA CAPTURE / TRANSFER

Extract / Recycle Manufacture / Process Stock Build

ROBOTIC & AUTONOMOUS PRODUCTION / TRANSPORTATION

ht Project Production Institute

The rapidly expanding fields of data science, machine learning, and artificial intelligence will have a major impact on society. OpenAI, through its research and suite of products; Alphabet (Google); IBM; and numerous other companies are researching, developing, and bringing to market products that allow users to rapidly write papers, create artwork and graphics, write music, and so on. But how will these rapidly emerging technologies affect the construction industry?

During a recent call for papers for an upcoming PPI technical conference, we see how professionals are looking to automate routine tasks such as cost estimating, specification writing, operating manual preparation, and maintenance scheduling, which are just a few areas where ML is being applied.

As construction becomes more industrialized and deploys autonomous machines, endless data will be created by these machines. Autonomous vehicles, robots, production machinery, and smart hand tools will take in and send out information at submillisecond speed.

A study by Intel, discussed by Intel CEO Brian Krzanich at Automobility LA, suggests that just one autonomous vehicle will generate around four thousand gigabytes (around four terabytes) of data *every day*. And that's just assuming one hour of driving! For more heavily used vehicles, it's expected that autonomous vehicles will generate and consume around forty thousand gigabytes (around forty terabytes) of data for every eight hours of driving time.[20]

Parallel technologies support this massive hoovering-up of data. Now we have hyperfast 5G; 6G is already under development, while 10G is being contemplated. Massive data centers serve as the new mainframes. Staggering amounts of fiber optics are being installed each day, while massive investment continues in support of quantum computing.

With large and ever-increasing volumes of data collected and created by today's organizations, how will all this data and information be used, and what will it be used for? How will it create value? Perhaps the most important question we almost never see being asked is "Who owns the data?" Here in Silicon Valley, data is what gold was to miners in the mid-1800s and what hydrocarbons were to energy companies at the turn of the last century.

---

20   Mewbern Ellis, "5G and autonomous vehicles—accelerating data communication speed," April 2020, accessed July 6, 2023, https://www.mewburn.com/news-insights/5g-and-autonomous-vehicles-accelerating-data-communication-speed.

Additionally, in the coming years the construction industry will learn how better to use photogrammetry and continue to find new applications for sensors and robotics. The use of mixed-reality technologies integrated with PPM solutions will more effectively design work processes and operations, allocate resources, and manage sources of variability.

Reality capture, IoT sensors, and autonomous vehicles will provide project production system performance feedback information while providing feed-forward instruction and optimizations. Artificial intelligence, machine learning, and robotic process automation will inform and automate key decisions.

Unfortunately, Era 1 and 2 frameworks based on productivity and predictability result in the following outcome: the construction industry is not able to leverage, much less *use*, all this data.

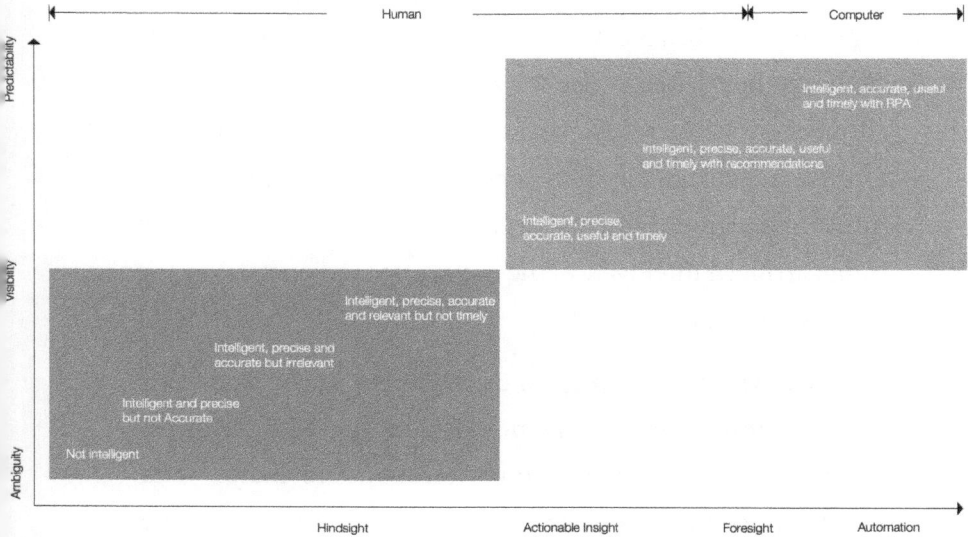

Copyright Strategic Project Solutions, Inc. – All Rights Reserved

WHAT IS THE ROLE OF OPERATIONS SCIENCE, DATA SCIENCE, ARTIFICIAL INTELLIGENCE, AND MACHINE LEARNING?

Eager to capitalize on fixing the construction industry, investors are seeking ways to apply AI/ML to the construction process. But Era 1 and Era 2 thinking comes back into play. The overreliance on administration and underinvestment in managing production makes return on investment related to AI/ML for construction a challenge. But adoption of the production-based perspective provides the framework for leveraging AI/ML, especially when integrated with operations science. In this case, operations science defines how the production system should behave and the implications of what will happen if the production system does not behave as it should. AI/ML can then be used to determine why it is not behaving or even predict what risks may make it not behave. Operations science also provides the basis for training the AI/ML.

To enable a better understanding of these technologies, OpenAI was asked to explain the difference between data science, machine learning, and artificial intelligence; to explain data science, machine learning, artificial intelligence, the metaverse, and operations science. The following is the response:

> **Data Science** is the process of extracting knowledge from large amounts of data. It includes the use of algorithms, statistics, and machine learning techniques to analyze data and extract meaningful insights. Data Science involves the application of scientific methods, processes, algorithms, and systems to gain insights from structured and unstructured data. This can involve using predictive analytics, data mining,

and machine learning to discover patterns and relationships in data. Data Science is used to gain insights from data and to solve complex problems. It is used in many areas, such as healthcare, finance, marketing, and more.

Machine Learning is the field of study which focuses on developing algorithms that allow computers to learn from data. It involves the development of algorithms that can recognize patterns in data and use them to make predictions or decisions. Machine Learning algorithms use statistical methods to find patterns in data, such as clustering and classification. These algorithms are then used to predict future outcomes or make decisions. Machine Learning is used in many areas, such as healthcare, finance, marketing, and more.

Artificial Intelligence (AI) is the field of study dealing with the development of machines that can think, act, and behave like humans. This involves the use of algorithms, machine learning, and deep learning techniques to create machines that can recognize patterns, make decisions, and solve problems. AI systems are designed to take in data, learn from it, and make decisions or predictions based on the data they receive. Examples of AI systems include self-driving cars, facial recognition systems, and virtual assistants. AI is used in many areas, such as healthcare, finance, marketing, and more.

Deep Learning is a subset of Artificial Intelligence (AI). It is a type of machine learning algorithm which is used to recognize patterns in data by using multiple layers of neural networks. Deep Learning algorithms are capable of learning complex patterns from large amounts of data. They are used in many areas, such as computer vision, natural language processing, and autonomous vehicles. Examples of deep learning applications include image recognition, speech recognition, and natural language processing. Deep Learning algorithms can be used to solve complex problems and generate insights from data.

The **metaverse** is an interconnected virtual world that exists in the form of a computer-simulated reality. The metaverse is made up of virtual environments, such as virtual worlds, augmented reality (AR) platforms, and mixed-reality (MR) applications. It is a shared online space where people can interact, create content, and participate in a variety of activities. The metaverse is a growing trend, as more and more people are turning to virtual spaces for social interaction, entertainment, and business.

Operations science is the study of how to improve and optimize processes and systems to achieve the desired objectives. It involves the use of mathematical models and other techniques to analyze and optimize systems. Operations science can be applied to many areas, such as manufacturing, logistics, supply chain management, and healthcare. It is used to improve efficiency and reduce costs, while ensuring that the quality of the output remains high. Operations science is used to improve the effectiveness of operations, while also reducing waste and improving customer satisfaction.

In a recent PPI presentation, Mark Spearman and Phil Kaminsky weighed in on the differences and applications of operations science and data science. They spoke about how both seek to explain how a complex world works and referenced IBM's definition of data science: "Data science is a multidisciplinary approach to extracting actionable insights from the large and ever-increasing volumes of data collected and created by today's organizations." They also explained operations science is the science that describes the behavior of operations. They then proposed that operations science be used to model, analyze, optimize, and better understand the production system, while data science should be used to analyze complex data, in real time, to gain insight into the production system's behavior.

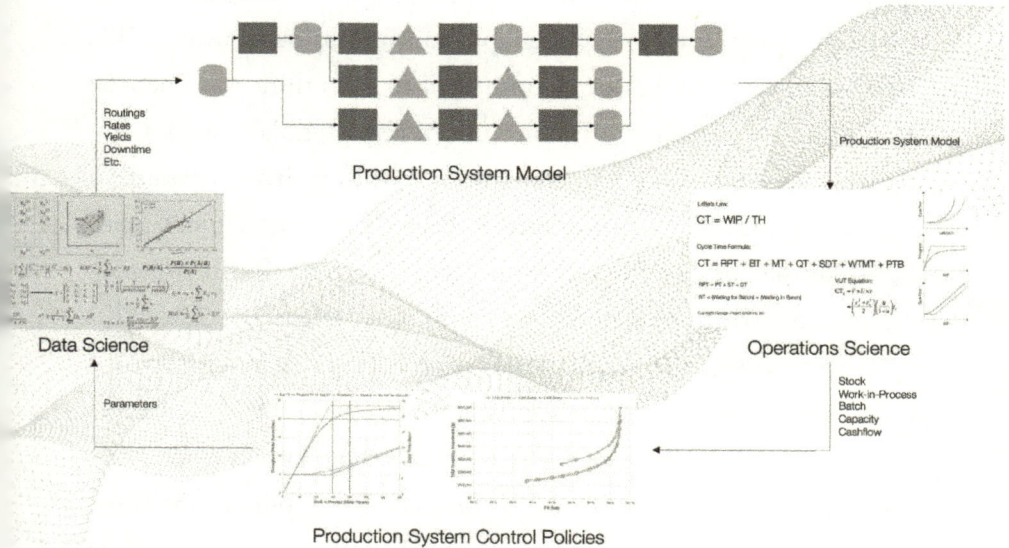

Production System Model

Data Science

Operations Science

Production System Control Policies

SELF-ORGANIZING, SELF-OPTIMIZING, SELF-CONTROLLING

We can see that these technologies will soon integrate together in one system, allowing for production systems to map, model, optimize and control themselves, setting the foundation for "self-forming/self-controlling production systems" or what, at SPS, we call Intelligent Production.

At SPS we are doing just that: enabling customers to compile this data into digital twins that facilitate operations science–based simulation, analysis, and control and facilitating optimization of any form of production system to supply chain. Then we combine this machine learning with operations science in a powerful one-two punch. Everything is connected, synchronized, and optimized. But this is just the start, as soon the modeling and optimization will also be automated.

In the past, production systems, including the related operations, routings, and performance metrics, needed to be mapped. Today, rather than trying to map it first, you can build a production system model that maps itself by picking up data from various sensors (GPS, RFID, IoT, trackers/tags, etc.) connected to both things and people. In this way, we can gather vast amounts of very granular data describing what is being worked on and what (or who) is sitting around.

Take a pipe spool, for example. There is no object more inert, dumb, simple, or devoid of information than a pipe. But if you stick a sensor on it, the pipe is now a data point in a system; the pipe tells a story. So maybe a pipe, or pipe spool, starts off as raw material with a sensor, but the sensor knows the pipe has been cut and had the flanges welded onto it, knows how long that process took, knows when it has moved from inbound raw materials or even the supplier to the shop, and knows it's at the cutting station, or the welding station, or the inspection station. It knows when it's outbound, transported to the site, off-loaded, staged, moved to the point of install, picked up by a crane, put in place, bolted up. And it might even know when it's been tested.

Thanks to the sensor, the computer can automatically pull all that information, synthesize it, and display the routing without us needing to map it manually. We can track the journey, including geography, time, and environment (e.g., temperature, which for some critical equipment may be important).

We can use this production system model of all these flows and use operations science to compare that to where we need to be and conduct simulations, analysis, and optimization.

None of this stuff is really that *complicated*, conceptually. But it *is* profound. It's going to change the way we work.

Similarly, we can measure the work done by workers. We know when a worker clocks in. If they're working on a refinery or data center, they need to go through a security point. And increasingly, as such technology becomes more ubiquitous and refined, we don't merely know that they've entered; we can pinpoint their exact location at a given time. There may be a security rationale for this kind of tracking, but we can use that data for production purposes.

> None of this stuff is really that *complicated*, conceptually. But it *is* profound. It's going to change the way we work.

This technology allows for a high degree of control through automation. It changes our thinking about what optimization means based on advanced operations science.

As I said in an earlier chapter, we're not just trying to make Schmidt work as hard as possible; we're trying to *maximize the flow through the process*. That was Taylor's blind spot, his fatal flaw. That's why operations science is critical. And the marriage of operations science with emergent technology creates exciting new vistas for the entire construction industry.

CHALLENGES OF PLUGGING CONSTRUCTION INTO THE INTERNET OF THINGS

This is powerful stuff. But every potential application brings challenges. For example, it's fun to talk about the miraculous things an IoT sensor can do on a pipe—but how do you attach it to the pipe? It's not as simple as it seems. Even before that, do you instrument the pipe, the pallet, or the transit vehicle? Some? All three?

What's the relationship between a pipe that is part of a final product going into an asset versus a truck that's already instrumented? The pipe's data is unidirectional—going from point A to point B (the point of installation)—but a truck is multidirectional, going from A to B and back to B again (or maybe with a detour to C ...)

How are the tags secured to the pipe, pallet, or truck? Glued? Welded? Are they permanent? Are they temporary? How do you protect their physical integrity in the rough-and-tumble world of construction?

How do you set up the network to take in the signal? And as you take it in, how do you clean up massive amounts of data to siphon out desired information from noise? Is the data "dumb"? Is it "intelligent" but late, or irrelevant?

> How do you set up the network to take in the signal? And as you take it in, how do you clean up massive amounts of data to siphon out desired information from noise?

Underlying these questions is a much bigger question: *Why would you do all this?* What do you do with all that data? How do you leverage it to produce something of value? There is a tendency to collect data just to do it—because it's the latest big thing, because everyone else is doing it, because it feels cutting-edge, etc. The tech industry is most guilty of this, but brick-and-mortar industries have followed suit, seeing the capability of data to do great things but often lacking a coherent strategy around it.

I know of a company that spent $100 million contracting a data science company. They sent the company a bunch of info, waited six months, called them up, and said, "What'd you guys find?" The response: "Nothing. What were we supposed to be looking for?"

We're in a transitional period where many people are doing "technology for the sake of technology," which is dumb, to put it bluntly. One big motivator is simply money: it's a way of cashing in. Business owners are enamored with the valuation multiplier associated with tech or tech companies. If your company is valued at one time or two times revenue or even profit, when it becomes a tech company or software company, the valuation is based on a much higher multiplier, up to thirty times revenue. So many companies are working diligently to *become* technology companies simply because it increases their valuation and becomes much more attractive to investors or as an acquisition target, while those with equity options see their personal wealth increase as well. In many cases, there is no real strategy or purpose behind it, however. Look at Tesla, for example. It's highly valued because it's regarded as a tech company. If it were simply classified as another auto manufacturer, its valuation would be lower.

Another problem is automating things that shouldn't be automated. Sometimes, the simpler, less-technological way of doing things is preferable. Another way of looking at this is through the distinction between automation and innovation, which are often conflated, but they are not the same. To automate means to apply technology to make what you possibly shouldn't be doing at all more efficient; innovation is trying to use tech to create a whole new way of doing something better. I've quoted it earlier in the book, but Peter Drucker's words of wisdom bear repeating: "There is nothing so useless as doing efficiently that which should not be done at all."

For example, a company applied AI to optimize onshore hydrocarbon drill rig movement. Once deployed, the solution worked as planned, but the resulting number of wells that were drilled—but not completed—also increased. The assumption was that if all operations within a process are optimized, the process itself will be optimized—Taylor's view. But they were wrong. As operations science dictates, the optimization of the drill rig just resulted in WIP building up between well drilling and well completion. They didn't actually get any more oil out of the ground. It was a well-intended action with unintended consequences. And you know what they say about the road to hell …

Technology automates and enables technical solutions, which should address specific business opportunities and problems. This may seem too obvious to even state, but believe me, there are many billion-dollar companies out there that need to hear the obvious, lest they make the same mistakes.

DIGITALIZATION AND PROJECT PRODUCTION MANAGEMENT

To understand the relationship between digital technology and project production management, we must accept that PPM is based on the concept that projects are production systems; hence, operations science can be applied to effectively understand and influence project outcomes. If an enterprise desires to achieve better project performance, focusing on how exactly work is planned and performed is a must. In this regard, PPM is used to map, model, analyze, simulate, optimize, control, and improve project production systems.

Today's digital technology naturally fits with PPM and its intent to optimize, control, and improve project production systems. Smart supply chains use IoT sensors to track location and environment for products; machine learning and artificial intelligence are used to understand and influence supply chain performance based on the data being captured.

These feedback and feed-forward loops will inform product catalogs that will be used to plug and play both products and processes. Cost estimating and scheduling will become fully automated, running in the background, while design teams make decisions and selections at each step of the process.

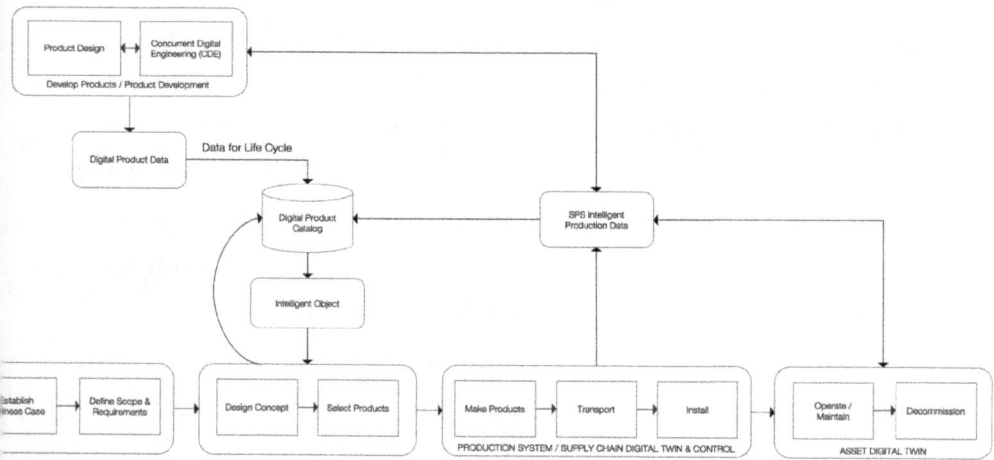

A DOUBLE-EDGED SWORD

It's exciting, really, but we shouldn't overlook legitimate concerns in our zeal to technologize our work.

Cinema fans remember the famous scene in *2001: A Space Odyssey*, when HAL 9000, the AI assisting the astronauts in space, turns against—overrides—its human handlers ("I'm sorry, Dave, I'm afraid I can't do that" is the memorable, and chilling, line). The film was pure sci-fi when it debuted in 1968, but half a century later, many of HAL's advanced computing capabilities, which were fanciful back then, are working technologies now.

Fortunately, I don't see robo-excavators on a jobsite becoming sentient before revolting against and killing their human overlords any time soon, but there will always be a tension between advanced technology and its human arbiters. Are we using the tech, or is it using us? As digital technology becomes more sophisticated and smarter, and increasingly seems to mimic aspects of human cognition, these concerns will be more pronounced. And construction is likely to be another battleground in the contest.

> There will always be a tension between advanced technology and its human arbiters.

Any time a disruptive or revolutionary technology emerges, usually you see three camps emerge in response. There is the person who tries to fight it (sometimes rationally, because the technology threatens to put them out of a job), the person who is indifferent, and the person who leverages it for their own benefit.

We should be in the latter camp, while acknowledging and mitigating the risks, drawbacks, and dangers that advanced technology presents.

CHAPTER 11

FRAMEWORK FOR TRANSFORMATION: HOW TO ENGAGE PEOPLE, INFLUENCE BEHAVIOR, AND ENABLE THE FUTURE OF CONSTRUCTION

WE CAN EFFECTIVELY APPLY all the lessons in this book, but there's one big bottleneck in every construction project and any initiative to improve—the human brain.

Nothing else matters if we can't actually (1) persuade key stakeholders that the changes we've discussed are worth implementing and (2) get them to change their behavior to do it. This is really a question of mental states, of dislodging entrenched ways of thinking and aligning current mental states with some vision of where we want to go.

This chapter will shift the focus somewhat, from the very large megaprojects we measure in number of years, in billions of dollars, in tons of materials, and thousands of workers to the small three pounds of squishy gray matter in your skull.

This chapter is where construction meets neurology/psychology. Without understanding that, everything we're talking about remains on the page, in the abstract. And our goal is to make it real.

THE SCIENCE OF CHANGE

What we are proposing can conflict with existing ways of doing business in the construction industry, so moving from a current state to a future state vision is not straightforward. Implementing these ideas has less to do with the concepts and techniques per se than it does with displacing an existing *philosophy*. It is really a business problem. And business problems are often psychological problems—figuring out what makes people tick and getting them to do what they're supposed to do, in concert with other stakeholders.

Everything starts, and ends, with the brain.

Over the past three decades, those of us at the PPI and Strategic Project Solutions have had the opportunity to work with several leading industrial psychologists, sociologists, and neuroscientists, among them Anil Seth, one of the most brilliant minds (no pun intended) in the field of neuroscience. Our collaboration with these thinkers has informed our approach to managing change and will ultimately, if adapted on a large scale, lead to better project outcomes worldwide.

We all like to think of ourselves as perfectly rational beings who make decisions based on a careful consideration of the facts at hand, but the truth is that there is a lot of unconscious processing, not all of it "rational," going on under the neurological hood. You may have heard the saying "People buy emotionally and justify the purchase rationally." Or as Tony Robbins puts it, "The cost of staying the same has to be greater than the pain and cost of the change—if it is not, there will be no effort to change."

When people learn something new—say, they read an eye-opening book, or take an illuminating course, or participate in a seminar where unconventional ideas are shared— they get excited in a positive way. The novelty of it elicits excitement, even exhilaration, in their mind—so much so that they often go and tell lots of people (colleagues, namely) about what they learned: "Guys, we've been doing things wrong the whole time!"

But they often find their colleagues don't share that enthusiasm. And why would they? They didn't read the book. They didn't attend the course. They skipped the seminar to play golf. They're just hearing all this new, revolutionary stuff second-hand. And most often by way of rapid-fire delivery.

The more you talk about your newfound knowledge, the less interested the other people become, to a point of confusion. Maybe you've encountered such a person in your company or among your friends (or maybe *you* were that person). "What's with Wayne? He can't stop talking about operations science. What's the deal?"

THE LADDER OF INFERENCE

Anil has taught us that the brain cannot hear, see, smell, feel, or taste. It can only receive electrical impulses from your eyes, nose, etc. To minimize effort and preserve energy, the brain is looking *for* it, not always *at* it. What this means is that the brain is perceiving, or as Dr. Seth states, "hallucinating." The brain takes shortcuts, makes assumptions based on past patterns and predictions based on incomplete data, and ignores certain stimuli in favor of what it deems important (all done unbeknownst to your conscious self, of course).

This cognitive system works well; the fact that eight billion humans are alive today is testament to the evolutionary adaptivity of the brain. However, it does have its shortcomings. One is that our perception is faulty. An excellent demonstration of this is optical illusions, a number of which you can find online. Some of the famous ones you've probably seen before. There was the dress controversy that took the internet by storm in 2015: half the people saw a dress that was black and blue, the other half saw white and gold. Each side was as sure in their identification of the correct color as they are sure that 2 + 2 = 4. Or the rabbit-duck illusion, which has been around in some form or another since the nineteenth century:

Welche Thiere gleichen einander am meisten?

Kaninchen und Ente.

Is it a rabbit or a duck? Both? Neither? The answer is in the eye of the beholder.

The ladder of inference helps explain the inner workings of our mind in the decision-making process and illuminates how our individual beliefs put a spin on how we see the world. Even cold, hard facts are processed through the lens of our individual experiences and particular worldview. This is why at a meeting, two people can look at the same set of data (or the same dress, as it were) and draw completely different conclusions about what they mean. Martin Fischer led some research related to how much time is spent on a construction site trying to understand and agree on what is being depicted on a set of drawings. It is pathetic that 2D drawings are still the norm, but it makes sense for the very reason Martin led the research. If we can't all align in what the drawing is attempting to communicate, then we have an out.

> The ladder of inference helps explain the inner workings of our mind in the decision-making process and illuminates how our individual beliefs put a spin on how we see the world.

The human brain is a powerful sensory data processing machine: at every given moment, it's taking in vast quantities of information about the world around it, and it must quickly and automatically filter out the signal from the noise: what is relevant to your ability to survive and thrive in your environment (signal) versus what is inconsequential (noise). For example, a new "thing" appears before you: that could mean a new hire at work, a new piece of information, or a bus hurtling down the street in your general direction. Is it a threat, a benefit, or something that can be ignored?

We synthesize this filtered data with our preexisting assumptions (which we've accumulated based on a lifetime of experience), and that leads to taking action. The actions and their consequences then *feed back into* the first rung of the ladder, and the process continues cyclically. It is self-reinforcing.

We see the world not as it *is* but as we believe it to be. Reality doesn't dictate perception; our perception dictates reality. The other day I saw a bumper sticker that humorously captures this idea: "Don't believe everything you think," a twist on the usual "Don't believe everything you see." Your thoughts might be errant; you may misinterpret environmental signals. It's hard for people to really accept this because our thoughts *feel* so real and so firm. Of course they do. That's how we interpret the world around us. As Anil says, we don't operate in a "I will believe when I see it" paradigm. Rather, it's "I will see it when I believe it." Take a moment and think about this.

It's counterintuitive, especially because most of us have an unshakable faith in ourselves as objective, rational beings, but that's not the whole picture. The operation of the human mind is fuzzier and more complex.

THE RETICULAR ACTIVATING SYSTEM

The reticular activating system (RAS) is a network of nerve fibers located in the brainstem that acts as a filter for incoming information. It determines which information reaches the conscious mind, and thus can influence what decisions are made.

The RAS filter can be difficult to get through, as it is designed to filter out incoming information that is irrelevant or not important. To get through the RAS filter, it is important to focus on the most important information and ignore the rest. Additionally, it is important to clearly communicate the purpose of the information, as well as how it will benefit the individual. Finally, it is important to be consistent in presenting the information, as the RAS filter is designed to recognize patterns. By following these steps, it should be possible to get through the RAS filter and make sure that the important information is received.

CHANGE IS SCARY

Although humans have a mild predilection for novelty, we are at the same time also hardwired to fear anything new. There is an evolutionary reason for this: our ancestors had to balance their desire to try new things, experiment, and explore with the harsh reality that taking an unfamiliar path through the forest might lead you off a cliff or eating berries from that unknown bush you came across might kill you, even if they look delicious.

We fear the unknown—change—because our DNA understands, in a very real sense, that the unknown can kill us.

In the modern world, particularly in the world of work, the fear of change becomes misaligned with the actual threat presented by that change—in other words, we often blow it all out of proportion. This explains why even good ideas get ignored or dismissed. They're threatening simply because they are new. That's just the brain doing what it does best: trying to protect you, for better or worse.

Sometimes this takes the form of missing the forest for the trees and failing to see the big picture. Take certain premises that are dominant in construction: "Through earned value, the more work I get done, the more progress I make" or "The harder I make people work [in a Taylorist sense], the more throughput there is." There is a kernel of truth in these ideas, but the problem is that they are counterproductive. They are localized approaches that interfere with the efficacy of the whole. All these lack systemic views of the system as a whole thing.

And underlying mental models reinforce these beliefs. We get stuck. We can't see the real object; all we see is the illusion.

To be fair, not all fearful reactions to change are purely irrational or knee jerk. Sometimes change *does* suggest a certain amount of risk to something other people hold dear, such as their job. People might implicitly understand that sweeping changes about how projects are carried out jeopardizes their current business model. I talked earlier about one reason why things *don't* change, even though everyone knows they're dysfunctional, is because the financial incentives for continuing them are too powerful. The cost of the change outweighs the cost of staying the same.

Later in this chapter, I'll talk about how to navigate these concerns in a way that recognizes the validity of people's feelings, while finding a way to lead them to embrace change and see it not just as a threat but as an opportunity.

THE CHEMICAL BRAIN

Now, let's get even more granular, digging down into the molecular level. Because when we talk about neurology, we're really talking about chemistry: how molecules interact to produce electrical and chemical signals that enable consciousness.

How we respond to new information, and how that response influences our decision-making framework, is to a great extent a function of the interplay of various chemicals (hormones, neurotransmitters, etc.) within your gray matter. These chemicals affect cognition, mood, behavior, and other mental functions, not to mention a whole bunch of physiological ones.

And how people *feel* affects what they *do*. That's why it's important to understand. Endorphins (a hormone) and dopamine (a neurotransmitter) are associated with an openness to change. These are (in simplified terms) "positive" chemicals that induce good feelings.

Serotonin (neurotransmitter) and oxytocin (hormone) are also positive. They are associated with feelings of togetherness and contentment. No wonder, therefore, that they play a role in the success of group activities.

Cortisol is the body's primary stress hormone, and one of its triggers is unfamiliarity. If a person in a ski mask jumps in front of you on a dark street late at night, that cortisol is gonna be pumping through you! Cortisol has utility: it primes the mind and body to handle threats. But again, sometimes it kicks in in settings where it is not helpful.

Anything perceived as unknown—and therefore potentially threatening—has the potential to elevate cortisol, which will make you feel emotionally and physically stressed. Obviously, stress inhibits the change process because it turns people against what you're doing.

Adrenaline has a similar function. It increases respiration and heart rate and primes your body for fight or flight in response to danger.

When I present at a conference, or confer with clients, I'm cognizant of how all these neurosubstances are at work underneath the surface. It's all neurochemistry. Adrenaline, cortisol (stress hormone, fight instinct).

For example, a while back, James Choo of SPS was facilitating a session at a prospective customer. The managers were gathered in the conference room. Toward the end of the presentation, the VP announced, "What's great about this is we'll no longer need scheduling!"

Right away James could spot who the scheduler was in the room, because he saw a woman in the corner shift uncomfortably in her seat and frown, looking worried. As it turned out, she reacted poorly to the (admittedly indelicately delivered) news because she thought she was going to lose her job. I've seen that a lot! Perhaps the most interesting experience in this area was a new hire on a large civil project. A VP was laying out the plan to transform the company. The new hire literally snapped and blurted out, "My mobile phone is crap, the car you gave me won't fit in the garage, and my dog is stressed out."

Once, someone in a large company said, "Good news is we can eliminate an entire department!" And that whole department, unbeknownst to the speaker, was sitting in that meeting. Needless to say, the cortisol and adrenaline pushed the resistance meter through the roof. You could see the people in the department thinking, "If we do this, I won't have a job, and without a job I can't pay my bills." The best was an executive who said, "I don't know what you offer, but whatever it is, I will resist!" At least this guy was honest.

When new information or models are presented to people who have something at stake, it stirs up certain chemicals, not all positive, in their brains. And this impacts behavior.

But what if there was a way to deliver the news in a way that elicits not just the molecules that provoke a fight-or-flight or hyper-vigilance response, but positive ones like serotonin and dopamine? The goal is to make them feel good about change!

LEAD THEM IN GENTLY: THE FOUR PHASES OF CHANGE

Change rarely comes easily. Mental models keep people stuck in their ways. The solution is to disrupt the ladder-of-inference cycle: get the brain moving, making new connections, form new neural pathways that lead to transformation.

Disrupt the Ladder of Inference

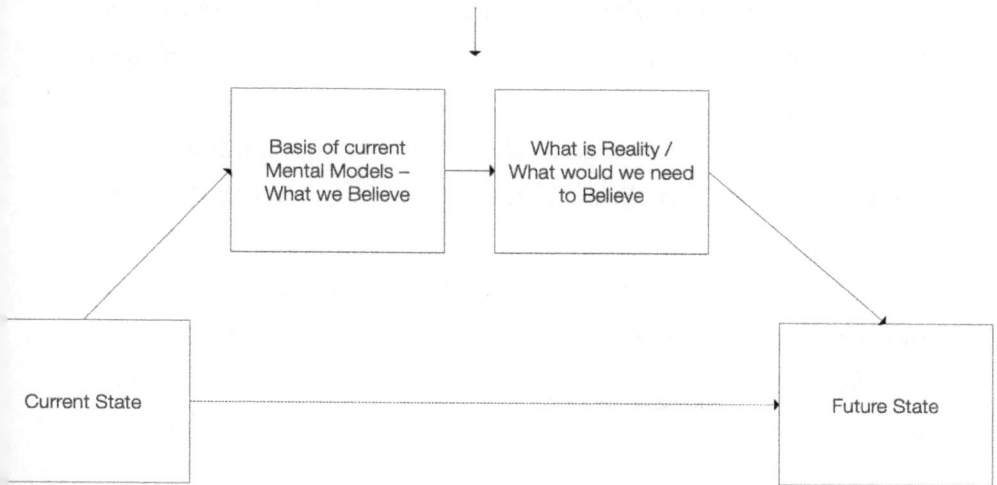

However, don't be misled by the word *disrupt*. It's not instantaneous; it requires patience and finesse. The four phases of change illustrate this process of moving from present state—being shown, and grappling with, new and perhaps contentious ideas—to a desired state of commitment, when the new ideas have been embraced.

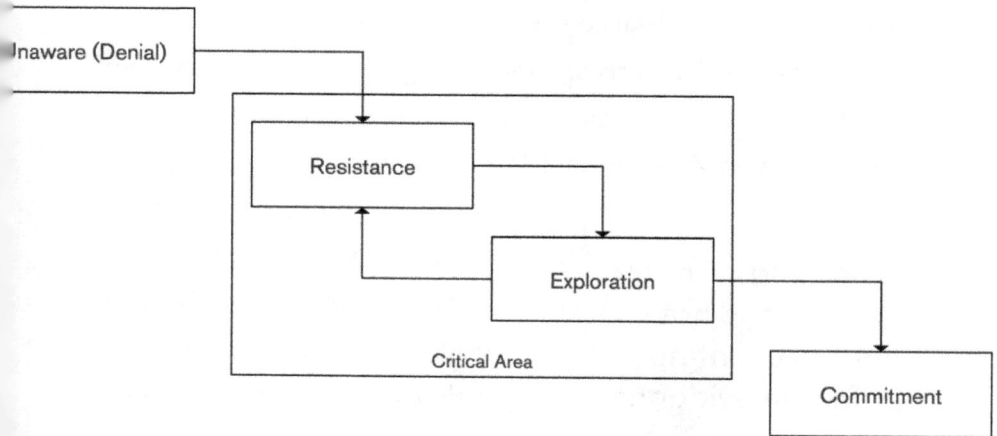

Simple enough in theory. But it can be a rocky road.

Your interaction with other parties or stakeholders must meet them where they are in a four-step process they must go through: unaware, resistance, exploration, and commitment.

A given interaction must be aligned to the situation. If in denial, inform; if in resistance, listen; if in exploration, guide; and if in commitment, support.

First, we have to inform the uninformed. Their reaction might be outright denial: "That will never work," "That doesn't make sense," etc. We often look to do interactive learning exercises that enable professionals to experience the various aspects of what we are endeavoring to have them learn.

Next, as we enter the critical area, they will go through a phase of resistance. Not only are they denying the validity of the new idea but actively pushing back: "That goes against everything I know. I can't be that wrong!" Or "That threatens my job/livelihood/profession / way of doing business!"

This phase can be very difficult for all parties. As Vic Ortiz (who works as an industrial psychologist and is a former SPS team member and well known among the lean construction community) says, the key is to "love the resistance—it says they are engaged."

Contrary to logic, navigating through this phase is *not* about presenting additional evidence. Again, humans are not purely rational beings; we make decisions and form judgments based on multiple, often conflicting, streams of information, not just who has the facts on their side. So dispositive evidence is often not effective to dislodge a long-held, but wrong or counterproductive, belief.

> **Navigating through this phase is not about presenting additional evidence.**

Over and over, we see people doubling down on explaining the concepts, theories, and potential value without understanding the other party's perspective, including how the ladder of inference and RAS may be in play.

That's why even showing people the results of past project failures *they have experienced themselves* is not necessarily adequate to dislodge their fixed belief.

We can't just browbeat them with facts and data. It will, as they say, fall on deaf ears. We have to apply a little finesse in transiting them through the phases of change. When we inform them, we must do so in a way that allows the lesson to sink in, to penetrate.

Remember, what you hear me say is hearsay; what you hear yourself say is fact! Imagine their minds are like a bucket filled with water. There's no room for new knowledge. You have to drain the bucket before you add water; otherwise, it's just going to continually overflow. Often, the best means of communication is questioning and probing versus telling and selling. We must suppress our natural tendency to tell and sell and take a softer approach, one that doesn't get cortisol and adrenaline pumping through the brain of your interlocutor.

Consider the following example—hypothetical but really not far off from real conversations I've had many times with various construction managers and owners.

"We have hired consultants to measure worker productivity," they say. Well, we know that's probably not a good idea—unless they're trying to optimize a bottleneck.

But we don't lead with telling them their idea is bad. Instead, we ask, "What do you hope to accomplish?"

"I need to get these fuckers to work!" is the response.

"OK, if they work, what might happen?"

And so on.

It's a little bit like the Socratic method used in classrooms, a little bit like the technique of talk therapists, who don't merely *tell* you what you need to hear, but ask reflective, open-ended questions that force you to scrutinize how you've done things and, by doing so, arrive at the answers yourself. We're trying to expose business models that reinforce counterproductive behavior and probe what it is they hope to accomplish.

We need to focus more on *questioning and collaboration* as a means to inform. Rather than telling and selling or demanding full and immediate compliance, we must be listening and learning and communicating the potential value of the methods we are proposing.

They might describe a problem they are having, one familiar to construction managers. "I brought the materials to the site, everything is ready, but I'm still behind schedule. How can that be?" Rather than saying, "Listen, you fucking idiot, it's obvious why. You should have done *XYZ*," we instead ask, "What did you hope to accomplish by bringing all the materials out there? How much are you spending on laydown on materials, just to have them lying around?" We take them through a journey but let them work through it from their own paradigm, which provides a bridge to understanding a new model.

A common response is something like, "That's the way we've always done it!" In that case, we must listen. If someone is resisting, we shouldn't simply double down on telling and selling. It is counterproductive. Listen, be patient, ask questions that lead them to the inevitable answer, and let them come to the conclusion themselves.

Resistance usually comes from three sources:

1. "It wasn't a good decision because I don't agree with the technical rationale behind it."

Let's have an off-site and jump off the Golden Gate Bridge and see who survives. Technically not a good idea.

2. "It wasn't the right process. There was a lack of input from or collaboration from others. You didn't involve the right stakeholders."

Once, I was in a meeting, and after one guy announced a decision, another guy said, "That gives me heartburn. Who was involved in that decision?" The first guy said, "Gary and I." To which the other said, "Oh, that's OK, as long as Gary signed off on it." Who is behind a company action has a lot to do with whether it's embraced or rejected.

3. "What's in it for me? Am I being treated with respect? Are they listening to my concerns?" The decision may have been a prudent technical decision, made by the right people, with the correct stakeholders involved, but if it goes against one's private interests, they will push back. Same if the process did not honestly honor and respect them.

Again, it's important not to simply dismiss these objections outright, as they may be valid. Nor is it the time to try to win by debating them into the dust or trying to persuade them using logical/rhetorical methods, the way an impassioned defense attorney would try to persuade a jury during closing arguments. Keep your mouth shut, take a step back, and just listen. Let them work through their own objections. Usually, the person will talk themselves into it and see your way.

To use another analogy, what do you do if your dog gets loose and you're running around the neighborhood trying to corral him? The more you chase, the more he runs away. But if you stop chasing him, the dog will probably come to you.

Eventually, the truth dawns on them—this is when you begin to enter the third phase: exploration. They see the change being discussed as an opportunity rather than a threat. As people work through resistance, they become curious about the possible rewards and/or benefits of change and begin to explore. A spirit of learning, experimentation, and possibility develops. For example, people in exploration seek new ways of doing things, begin to create a vision of the possibilities of the future, take risks, generate lots of ideas in support of the transformation process, and accomplish intermediate goals and celebrate milestones.

Resistance and exploration are closely intertwined and cyclical. The phase change doesn't happen all at once; there will be a lot of back and forth. This is the liminal period when the old paradigms start to fade away and the new ones crystallize in their place.

You can win someone's buy-in, but without proper guidance to nurture the exploration that is percolating and solidify it into actual transition, they may just revert back to the same old patterns, recreating the same mistakes based on underlying mental models. The key is to find ways to reshape the neuropathways.

I've seen it many times: companies try to implement new means of doing something, they put together a team to execute it, and they make some progress or start breaking through, but people end up reverting to status quo.

Finally, once you get commitment, you must support them and reward them for doing what they were supposed to do. It would be great if just leading them to the aha, lightbulb-above-the-head moment was the end point. But getting someone on board with a new methodology is more a process than a moment.

DOS AND DON'TS

Implementation is a science unto itself. These are a few dos and don'ts for the process of breaking through entrenched mental models.

"Don't hold any barbecues": This happens a lot, and not only in construction. A company decides they're going to implement some bold new strategy. They create a catchy slogan and hold a party to celebrate it. Something like "We're going lean!" or "We're a digital enterprise now!" Sounds nice. But people see through the bullshit. They're sitting there eating their smoked brisket while saying, "Well, fine, but it doesn't look any different from before." Same nonsense but with a new name. And they're right.

And right away you lose people and force them into resistance mode. Drain the water first before you start throwing parties.

Get your shit straight: What, specifically, are you actually proposing/doing? You must be clear and concise. No bullshit, no filler, no "glittering generalities" (empty phraseology that sounds nice but means nothing), no vague statements. Remember, RAS is always filtering. Write a concise manifesto articulating what you want to change and how you intend to do it. Do this collaboratively. Bring in other stakeholders. Seek input and feedback, including from subordinates. This way, they have ownership of the process and also learn about what you're trying to do before you even launch the program.

Be aware of what neurochemicals you are triggering with your program: Happy chemicals or angry chemicals?

If you say, "We're downsizing this department," well, you're going to get resistance (adrenaline, cortisol, stress hormones). On the other hand, effective collaboration and team building triggers serotonin and oxytocin, and the thrill of novelty or solving a puzzle can stimulate dopamine. Business is nothing but a group of humans engaged in some collaborative, profit-seeking task. Humans are emotional creatures, and emotion is influenced by the cocktail of hormones and neurotransmitters sloshing around in your head. At the risk of over-simplifying, business equals brain chemicals. Understand the latter, and you'll excel at the former.

Articulate what you are going to *stop* doing in addition to what you want to *start*.

If you're leading a change campaign, you must be clear about three things: which new activities we will *start* doing, which current activities we will *keep* doing, and which current activities we will *stop*.

Each project is an opportunity for growth and experimentation.

Because of its project-based nature, construction is an ideal laboratory for applying new methods and facilitating some kind of transformation. It is more ephemeral and dynamic than other industries. In manufacturing you might be stymied by rigid supply chains and fixed, heavy equipment bolted to the floor—you can't transform a factory overnight.

But construction is, at its essence, a series of projects. Each project provides an opportunity to try new things and apply novel techniques and knowledge.

Rather than a more top-down implementation, like holding a company barbecue and making an announcement issued from the C-suite, we can use individual projects to create capability in a horizontal, peer-to-peer fashion.

> **But construction is, at its essence, a series of projects. Each project provides an opportunity to try new things and apply novel techniques and knowledge.**

Remember, PMs do *not* like top-down directives. One PM told me he got into his field because it was insulated from micromanaging, meddling upper-level managers. Out in the field, they leave you alone. Let us take advantage of that circumstance.

Getting involved in organizations such as the Project Production Institute is also beneficial. You don't have to go it alone. Learn from others whose journey is underway. Learn from their mistakes and be inspired by their successes. Share knowledge. Incubate collective wisdom.

LOOKING FORWARD

Between now and 2060, the world's population will double, and all those people are going to need structures where they can live, work, play, travel, and thrive as a species, from housing to office buildings to factories to entertainment complexes to the vital infrastructure that enables all of it. Just the amount of building floor space required will be equivalent to building an entire New York City every month for the next forty years.

But the construction industry is woefully underprepared for this. Hell, it's not even prepared now! As we've said many times throughout the book, construction is broken: it's built on shaky foundations and stuck in inefficient, old-fashioned ways of getting things done.

To meet these challenges, real change is required. Improved productivity of the construction industry is no longer just optional—it's imperative.

The time for business as usual is rapidly closing. The pain of the status quo in construction is going to increase exponentially as our capacity to develop and execute projects falls short of expectations.

Until we recognize projects as production systems and use operations science to drive project results, we are doomed to failure. We need to free ourselves from the prior eras and instead focus on a new era of project delivery, one in which projects will be highly efficient production systems that utilize the bounty of the technology (AI, robotics, data analytics, etc.) we are privileged to have access to.

AI-enabled systems will be ordering materials directly from suppliers without human hands. Facilities will be configured to order. Components will be 3D printed on demand at site. Workers with exoskeletons will be used to assemble components designed in completely different ways from today's methods.

But real change is not something someone else is going to enable. And while we can borrow techniques from manufacturing, tech, and other industries, that change will not come from elsewhere. It comes from us. We must lead the charge.

I promise that ultimately you will be rewarded for having the courage to try.